Mechanisms and Regulation of Carbohydrate Transport in Bacteria

Mechanisms and Regulation of Carbohydrate Transport in Bacteria

Milton H. Saier, Jr.
Department of Biology
The John Muir College
University of California, San Diego
La Jolla, California

1985

ACADEMIC PRESS, INC.
(Harcourt Brace Jovanovich, Publishers)
Orlando   San Diego   New York   London
Toronto   Montreal   Sydney   Tokyo

ACADEMIC PRESS, INC.
Orlando, Florida 32887

United Kingdom Edition published by
ACADEMIC PRESS INC. (LONDON) LTD.
24–28 Oval Road, London NW1 7DX

Library of Congress Cataloging in Publication Data

Saier, Milton H.
    Mechanisms and regulation of carbohydrate transport
in bacteria.

    Bibliography: p.
    Includes index.
    1. Bacteria––Physiology. 2. Microbial metabolism.
3. Carbohydrates––Metabolism. 4. Biological transport––
Regulation. I. Title.
QR92.C3S25 1985   589.9'019248   84–16949
ISBN 0–12–614780–9 (alk. paper)

PRINTED IN THE UNITED STATES OF AMERICA

85 86 87 88   9 8 7 6 5 4 3 2 1

To my parents, Milton and Lucelia Saier, on the
occasion of their fiftieth wedding anniversary

*Man's mind stretched to a new idea never goes back
to its original dimensions.*

Oliver Wendell Holmes

Contents

Preface

In 1980, we published a fairly detailed review in *Microbiological Reviews* entitled Carbohydrate Transport in Bacteria (Dills *et al.*, 1980). This review article, completed at the end of 1979, summarized and evaluated the literature then available concerning the mechanisms and regulation of carbohydrate transport in prokaryotic organisms. The present monograph was initiated with the intent of updating that review. However, within the elapsing five-year period so much progress has been made in understanding transport and its regulation at the molecular level that a review article of normal length could not possibly provide an in-depth analysis of all of the new information.

In this volume we shall see (1) that important structural and topographical features of several sugar permeases in *Escherichia coli* have been elucidated; (2) that the details of the energy-coupling processes have been clearly delineated for most (but not all) types of bacterial carbohydrate permease systems; and (3) that mechanistic details of the translocation processes have now been proposed. The genes encoding representative transport proteins within each major class of permease, those specific for lactose, melibiose, mannitol, maltose, and probably glycerol, have all been cloned, and at the time this volume was prepared, the nucleotide sequences of the *lacY* and *mtlA* genes encoding the lactose and mannitol permeases had been sequenced and analyzed. These two permeases have been purified to homogeneity and their vectorial reactions have been characterized in proteoliposomes consisting solely of phospholipids and a single transport protein. It can therefore be considered established that both of these permease proteins catalyze sugar transport in the absence of any other protein constituent of the cell. Mechanistic features of the two permeases have come to light as a result of combined kinetic, biochemical, genetic, and biophysical approaches. Comparable information is now available concerning outer membrane porins. In the first three chapters this information, together with the structural advances, will be evaluated.

Great strides have been made in understanding the mechanisms by which carbohydrate uptake and efflux are regulated. Five years ago, a specific model by which the phosphotransferase system regulates other uptake systems was available, but no direct biochemical evidence substantiated this model. The recent advances of molecular genetics and the more refined biochemical technology now available have allowed the establishment of this model as detailed in Chapter 4 and have revealed some unexpected features of the regulatory process. This system appears to be the best-characterized example of macromolecular mediation of information transfer from the external cell surface to multiple targets inside the cell.

The discussion in Chapters 4 and 5 also reveals that while the existence of several additional regulatory processes is well established, their mechanistic details are less well defined. In some cases, however, specific postulates have been put forth to explain these processes, and substantial evidence has been presented in their support. Since bioelectric, chemical, and macromolecular regulatory mechanisms are evidently operative, a number of parallels can be drawn with analogous processes in higher eukaryotes.

Perhaps the most startling advance over the last five years was the discovery of the unanticipated involvement of metabolite-activated protein kinases in the regulation of transport and the accumulation of cytoplasmic inducers in gram-positive bacteria. These advances, as well as the current literature on the involvement of similar catalytic agents in carbon metabolic control in gram-negative bacteria, are reviewed in Chapter 6. It is proposed that cyclic AMP and protein kinases first evolved to coordinate carbohydrate metabolism in prokaryotes and that subsequent diversification of these regulatory processes to more complicated processes in eukaryotes followed.

Not all inducers of carbohydrate permeases in gram-negative bacteria act from cytoplasmic locations. Some act from the periplasmic space in processes which are apparently mediated by integral transmembrane signaling devices. Both catalytic and structural models have been proposed, and the application of genetic engineering approaches already well under way should yield definitive information concerning these mechanisms in the near future (Chapter 7). Knowledge of these mechanisms may provide evidence concerning the evolutionary origins of transmembrane signaling processes such as those mediated by hormone receptors, cell adhesion macromolecules, and immunoglobulin receptors in animal cells.

Finally, in Chapter 8, an attempt is made to explain the diversity of transport systems throughout the living world and to point out the probable structural, functional, and evolutionary relationships of these systems to one another. The known carbohydrate transport systems in enteric bacteria are also classified according to mechanism. A survey of the list of permeases compared to the list of exogenous carbon sources which can be utilized by *Escherichia coli* and *Salmonella typhimurium* suggests that most of the carbohydrate transport systems in these organisms have already been characterized. It is probable that although considerable variation within each of the five principal classes of transport processes discussed in Chapters 2 and 3 will emerge, no fundamentally new mechanisms will come to light. A major task of the microbial biochemist is, therefore, to integrate our newly obtained structural information with functional studies aimed at defining the solute translocation mechanisms which serve as molecular gates separating the compartments of the living and the nonliving.

Milton Saier

Acknowledgments

I wish to thank the following colleagues for valuable discussions, for permission to cite unpublished results, and for suggestions regarding this monograph:

A. Apperson	M. Hofnung
K. Basu	G. R. Jacobson
P. Bavoil	H. R. Kaback
J. Beckwith	W. W. Kay
K. Beyreuther	J. Kyte
J. Brass	C. A. Lee
A. Danchin	J. E. Leonard
J. Desai	J. London
J. Deutscher	M. Neuhaus
S. Dills	M. J. Newman
R. Doolittle	R. A. Nicholas
B. Erni	H. Nikaido
S. Froshauer	M. J. Novotny
P. Gay	T. Osumi
S. Ghosh	M. Pfahl
E. Gilson	D. Printz
F. C. Grenier	J. Reizer
J. L. Guan	H. V. Rickenberg
W. Hengstenberg	G. T. Robillard

H. Rosenberg	G. Tenn
J. Rosenbusch	J. Thompson
H. Schindler	R. Tuttle
M. Schwartz	J. Y. J. Wang
I. Stuiver	E. B. Waygood
L. E. Tanney	T. H. Wilson

Particular thanks are extended to Leanne Sorenson for invaluable assistance in the preparation of this manuscript and Lynn Green, who drew the illustrations. Work in the author's laboratory was supported by NIH Grants 2-RO1-AM21994-0401 and 5-RO1-AI14176-07MBC.

Mechanisms and Regulation of Carbohydrate Transport in Bacteria

1

Introduction

Carbohydrates are transported into bacterial cells by a variety of mechanisms. Within the last 4 years tremendous progress has been made in defining these mechanisms and characterizing the proteins which catalyze the vectorial reactions. Conceptually, these proteins are thought to function by either of two general processes as shown in Fig. 1.1. They may translocate their hydrophilic solutes through aqueous pores (Fig. 1.1, I), or they may transport the solute by a carrier-type mechanism (Fig. 1.1, II). It is presumed that in both cases all or part of the hydration shell of the solute is lost as it passes through the membrane but that the hydration shell is restored upon re-entry into the aqueous solution on the other side of the membrane. While a pore-type mechanism may involve zero, one, two, or more solute binding sites, a carrier-type mechanism usually invokes

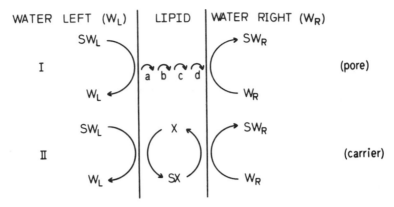

Fig. 1.1. Schematic depiction of two fundamentally different transport processes. Solute (S) with its hydration shell (W) becomes associated with a membrane protein which replaces all or part of the hydration shell. The molecule then passes through the membrane (I) via a pore-type mechanism in which the solute passes from one complexation site of the protein to another before re-entry into the aqueous environment on the opposite side of the membrane, or (II) by a carrier-type mechanism in which the solute passes through the hydrophobic barrier of the membrane complexed with a moiety of the carrier protein. The fundamental difference lies in the fact that only in the latter case does part of the protein shuttle across the membrane with the solute. Transport mechanisms can be envisaged which incorporate elements of both mechanisms. [Adapted from Mitchell (1979).]

a single site, which can shuttle between states localized to the two sides of the membrane.

In order to enter the cytoplasm of the gram-negative bacterial cell, an extracellular hydrophilic molecule must pass through both the outer and the inner membranes of the cell envelope (Fig. 1.2). Passage through the outer membrane is normally accomplished via hydrophilic proteinaceous pores. The properties of the best characterized carbohydrate-translocating porins in *Escherichia coli* are summarized in Table 1.1 (Lugtenberg and Van Alphen, 1983). Among these proteins are the *ompF* and *ompC* porins, which are fairly nonspecific with respect to their substrate specificities. Oligosaccharides of less than 500 daltons can generally pass through these pores, and amino acids, phosphorylated carbohydrates, and salts are also effectively transported. Synthesis of these proteins is dependent on the osmolarity of the growth medium. By contrast, the *phoE*, *tsx* and *lamB* porins exhibit substrate specificity, preferentially transporting inorganic phosphate and phosphorylated carbo-

TABLE 1.1

Characteristics of Outer Membrane Transport Proteins in *E. coli*

Gene and protein designation	Specificity	Induction[a] by	M.W.	Number copies per cell	Gene location (min)	Pore diameter (nm)	Oligo- meric form
ompF porin	General	Low osmolarity	37,205	$\leq 10^5$	21	1.4	Trimer
ompC porin	General	High osmolarity	38,306	$\leq 10^5$	47	1.3	Trimer
phoE porin	Phosphorylated compounds	P_i limitation	36,782	$\leq 10^5$	6	1.2	—[b]
tsx porin	Nucleosides	Nucleosides	26,000	$\leq 4 \times 10^4$	9	—[b]	—[b]
lamB porin	Maltodextrins	Maltodextrins	47,392	$\leq 10^5$	91	1.6	Trimer

[a] The *ompF*, *tsx*, and *lamB* porins have been reported to be subject to catabolite repression, emphasizing the roles of these channel proteins in carbohydrate transport across the outer membrane.
[b] —, Not known.

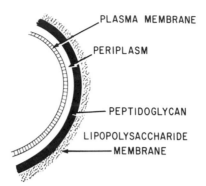

Fig. 1.2. Schematic depiction of the cell envelope of a gram-negative bacterial cell. The figure depicts the two membranes, the inner or plasma membrane which houses the energy-dependent carbohydrate-specific permeases shown in Fig. 2.1 as well as the outer or lipopolysaccharide-containing membrane which houses the porin proteins which are relatively nonspecific with respect to their substrates (Table 1.1). The peptidoglycan layer is presumed to be porous, so that small hydrophilic molecules can easily diffuse from the outer membrane across the periplasmic space to the cytoplasmic membrane. [From Saier (1979).]

hydrates, nucleosides, and maltodextrins, respectively. The specificities of these three porins presumably result from the presence of specific solute-binding sites within the pores, i.e., highly stereospecific arrangements of the aminoacyl residues, which comprise the pores. Induction characteristics of these proteins are in accord with the specificity studies, both of which suggest a specialized function for each protein (Table 1.1). The *lamB* porin (maltoporin) will be discussed at length in Chapter 2, Section D as it comprises part of the maltodextrin transport system. Other outer membrane proteins, which function in the transport of vitamin B_{12} and complexed ferric ions, have been identified (Lugtenberg and Van Alphen, 1983).

Recently the amino acid sequences of several of the major outer membrane proteins of *E. coli* have been examined for sequence homology (Nikaido and Wu, 1984). These proteins included the *ompF*, *ompC*, and *phoE* porins as well as the *lamB* (*malM*) porin, which encodes maltoporin (see Table 1.1). Significant local homology was observed for these proteins as well as for two other outer membrane proteins, those encoded by the *ompA* and *tolC* genes. These findings either suggest a common evolutionary origin for these different proteins or reflect a common mechanism by which these proteins are exported to the outer membrane.

By contrast with the transport proteins of the outer membrane, those in the inner, cytoplasmic membrane (Fig. 1.2) generally function by stereospecific mechanisms exhibiting a high degree of specificity as expected for a catalytic protein. These distinct transport processes are depicted in Fig. 2.1, in Chapter 2. The one exception to this rule is the glycerol permease. Glycerol has been shown to enter the *E. coli* cell by facilitated diffusion through a nonstereospecific hydrophilic pore of inner diameter equal to about 0.4 nm. The kinetic characteristics of glycerol transport have been well defined (Chapter 2, Section A). The lactose (*lac*) permease protein (Chapter 2, Section B), which catalyzes lactose: H^+ symport, has been solubilized from the membrane, purified to homogeneity, and reconstituted in artificial phospholipid membranes. The gene encoding this protein has been cloned and sequenced so that the primary amino acid sequence of the *lac* permease is known. Kinetic and chemical studies have led to postulates regarding the translocation process, and evidence is emerging that this protein functions by a carrier-type mechanism. Such a postulated mechanism involves a shuttling of the substrate-binding site of the permease between two states, each accessible to only one side of the membrane (Fig. 1.1). A distinct system, similar in several respects to the lactose permease, is the melibiose (*mel*) permease. This transport system catalyzes sugar: Na^+ symport, but it is also capable of catalyzing sugar: H^+ and sugar: Li^+ symport (Chapter 2, Section C). The activity of this protein has also been reconstituted in an artificial membrane. It appears to function by a mechanism analogous to that of the *lac* carrier, but with certain interesting kinetic differences. The tertiary structure of this protein probably resembles that of the lactose permease.

In contrast to the other systems discussed in this volume, maltose and maltodextrins cross the cytoplasmic membrane by a high-affinity active transport system whose activity depends on the integrity of four distinct proteins (Chapter 2, Section D). A channel-mediated translocation process seems likely for this multicomponent system. In order to ensure efficient scavenging activity at low maltodextrin concentrations, the bacteria synthesize a proteinaceous hydrophilic pore in the outer membrane (maltoporin), which, together with the periplasmic maltose-binding protein, functions to translocate malto-oligosaccharides across this structure. Much is now known about the structure and function of maltoporin.

Finally, like the *lac* permease, the mannitol permease (the mannitol Enzyme II of the phosphotransferase system) has been solubilized from the membrane, it has been purified to homogeneity, and its transport function has been reconstituted in artificial proteoliposomes (Chapter 3, Section E). The soluble energy coupling proteins of the system have also been purified and subjected to critical analyses (Chapter 3, Sections C and D). Detailed biochemical and genetic experiments have resulted in the identification of most of the PTS proteins (Chapter 3, Sections A and B). The gene encoding the mannitol Enzyme II protein has been cloned and sequenced so that the primary amino acid sequence of the protein is known. The mannitol Enzyme II and the lactose permease are therefore the two bacterial carbohydrate transport systems for which the structures and transport functions are best understood.

As a consequence of extensive biochemical and genetic analyses performed on the proteins of the phosphotransferase system, the structures of these proteins have been correlated with function (Fig. 1.3). Group translocation of a sugar (mannitol or glucitol in Fig. 1.3) probably involves five sequential phosphoryl transfer reactions and four distinct protein phosphorylation sites. All such sites involve histidyl residues in the phosphocarrier proteins of the PTS. The first two phosphorylation sites are associated with the general energy coupling proteins of the PTS, Enzyme I, and HPr. The second two sites are associated either with a single Enzyme II (as in the case of the mannitol system) or with an Enzyme II–III pair (as in the case of the glucitol system).

In the last section of Chapter 3, the possible evolutionary origins of the complex PTS found in enteric bacteria will be discussed. Evidence will be presented that all of the PTS proteins arose from a single ancestral protein, which may have catalyzed the phosphorylation and transport of fructose (Fig. 1.4). This unifying concept should provide the basis for correlating and coordinating many independent lines of PTS research with widely divergent bacterial species.

As transmembrane permeation represents the first step in the catabolism of exogenous sugars, one would expect the process to be subject to stringent regulatory control. Since the end products of sugar metabolism are utilizable sources of intracellular carbon and energy, the ideal situation would be one in which the permeases could sense the energy states and metabolite levels of the cell. These agents would exert negative control over permease function by

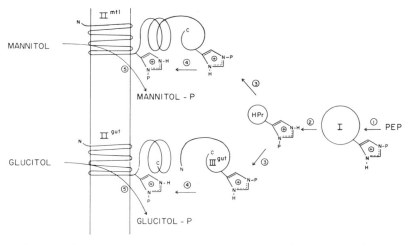

Fig. 1.3. Sequential protein phosphorylation reactions of the phosphoryl transfer chains for mannitol (mtl) and glucitol (gut) in *E. coli*. The figure shows five phosphoryl transfer reactions, which result in the phosphorylation of four distinct protein sites. Phosphoenolpyruvate (PEP) phosphorylates Enzyme I on the N-3 position of a histidyl residue, HPr is phosphorylated on the N-1 position of a histidyl residue; the first phosphorylation site on the Enzyme II^{mtl} and the phosphorylation site on III^{gut} are N-3 histidyl residues, while the second phosphorylation site on the Enzyme II^{mtl} and the phosphorylation site on Enzyme II^{gut} are thought to be N-1 histidyl residues. The last step involves transfer of the phosphoryl moiety from the N-1 position of a histidyl residue in the Enzyme II to the incoming sugar. The Enzyme II^{gut}–III^{gut} pair is shown to be structurally and functionally equivalent to the Enzyme II^{mtl}. It is proposed that all Enzymes II and Enzyme II–III pairs of the PTS exhibit similar structural features, involve the same sequence of phosphoryl transfer reactions, and catalyze sugar transport by essentially the same mechanism as illustrated in the figure. These speculative proposals result from their common evolutionary ancestry.

mechanisms which would be analogous to feedback inhibition of enzyme action. They would prevent the uptake of carbohydrates in excess of the needs of the cell. Five such regulatory mechanisms controlling transport function have been established in enteric bacteria (Fig. 1.5). Many permeases are, in fact, sensitive to the chemiosmotic energy state of the cell [the membrane potential (Chapter 4, Section A)] as well as to intracellular sugar metabolites including sugar phosphates (Chapter 4, Section B).

In addition to feedback inhibition of carbohydrate uptake by intracellular sources of carbon and energy, the cell would benefit from mechanisms which allow it to set up a hierarchy of preferred carbon sources. To a first approximation, a preferred carbon source would

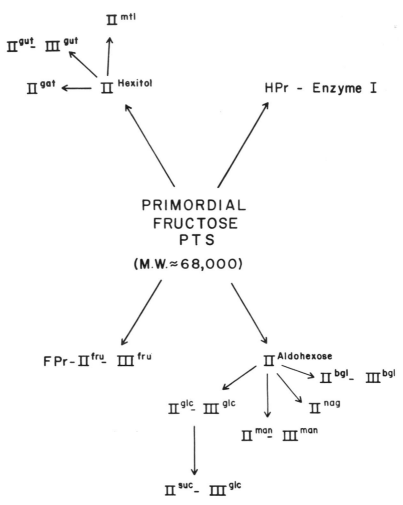

Fig. 1.4. Proposed pathway for the evolution of the enzymatic constituents of the phosphotransferase system in *E. coli* and *S. typhimurium*. This scheme is based primarily on the apparent relatedness of the Enzymes II of the *E. coli* PTS. The scheme indicates directionality in the evolutionary process with a fructose PTS (middle) as the primordial system giving rise to the present day fructose-specific system (bottom left), to the energy-coupling proteins of the PTS, Enzyme I, and HPr (top right), to the hexose-specific systems (bottom right), and to the hexitol-specific systems (top left).

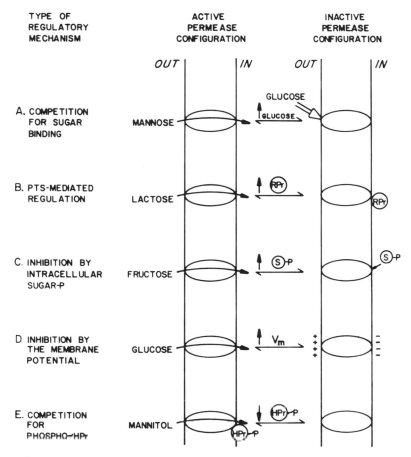

Fig. 1.5. Proposed carbohydrate transport regulatory mechanisms in bacteria. Representative examples occur in *E. coli*. Abbreviations: HPr, histidine-containing phosphocarrier protein of the PTS; PTS, phosphoenolpyruvate:sugar phosphotransferase system; RPr, regulatory protein (glucose Enzyme III [Enzyme III^{glc}] of the PTS); sugar-P, sugar phosphate; V_m, membrane potential or possibly the proton electrochemical gradient; S, substrate. [From Dills *et al.* (1980), with permission.]

be one which is readily available and feeds into a major catabolic pathway after a minimal number of metabolic conversions which are catalyzed by inducible enzymes. According to this assumption, glucose is preferred to glycerol or lactose in *E. coli* because the former sugar is concomitantly transported and phosphorylated in a process catalyzed by the PEP-dependent phosphotransferase system, and

futher metabolism is catalyzed by constitutively synthesized enzymes. Lactose and glycerol enter the cell by mechanisms which do
not couple phosphorylation to transport, and several metabolic
steps are required before these carbohydrates feed into the glycolytic pathway. Thus, glucose should regulate the rates of glycerol
and lactose uptake by a nonreciprocal mechanism. In Chapter 4,
Section C, we shall see that this is, in fact, the case (Fig. 1.5). The
detailed molecular mechanism by which the presence of extracellular glucose inhibits the activities of carbohydrate permeases which
are responsible for the uptake of less preferable carbon sources will
be discussed.

Because its conversion to fructose 1,6-diphosphate is catalyzed
by constitutively synthesized enzymes, glucose is also preferred to
other sugar substrates of the phosphotransferase system such as
mannitol and glucitol. A distinct mechanism for setting up a hierarchy of preferred PTS substrates has recently been recognized
(Chapter 4, Section D). This mechanism involves competition for
the common phosphoryl carrier protein of the PTS, phospho-HPr.
The preferred carbon sources are those which are transported by the
predominant sugar-specific permeases and possess high affinity for
phospho-HPr. Because the basal level of the glucose-specific system
is relatively high and because this system apparently has high affinity for phospho-HPr, glucose is utilized preferentially to other sugar
substrates of the PTS. Preliminary results in gram-positive bacteria
also suggest that protein kinase-catalyzed, ATP-dependent phosphorylation of a seryl residue in HPr can selectively inhibit utilization of one PTS sugar relative to another and that PTS permeases
themselves may also be allosterically regulated (Chapter 6, Section
C).

The product of the carbohydrate transport process, or a primary
metabolite derived from it, usually serves as an inducer of the sugar
catabolic enzyme system. The inhibition of sugar uptake by any one
of the mechanisms discussed in Chapter 4 is therefore referred to as
"inducer exclusion." In a previous article (Saier, 1979), we proposed that each of the mechanisms responsible for the inhibiton of
inducer uptake might also inhibit the cyclic AMP biosynthetic enzyme, adenylate cyclase. Coordinate regulation of cytoplasmic inducer and cyclic AMP concentrations would allow dual control over
catabolic enzyme synthesis. Recent work has, in fact, confirmed this
postulate. Dissipation of the proton electrochemical gradient has

been shown to inhibit the activity of adenylate cyclase, and high internal concentrations of sugar phosphates likewise depress the activity of the enzyme (Chapter 5, Sections A and B). Moreover, it has long been recognized that the phosphotransferase system regulates adenylate cyclase activity coordinately with inducer uptake (Chapter 5, Section C). It is also possible that energy competition provides a mechanism for the regulation of adenylate cyclase activity, but this possibility has not yet been examined. In any case, coordinate regulation of adenylate cyclase and the permeases responsible for inducer uptake by at least three distinct energy-sensing mechanisms appears to be established.

In recent years, it has become clear that sugar efflux as well as the uptake process can be subject to regulation. This process has been termed *inducer expulsion* (Chapter 6). When uptake followed by expulsion is allowed to proceed uninhibited, a "futile cycle" of energy expenditure occurs. Such a process has been demonstrated both in gram-negative and gram-positive bacteria (Chapter 6, Section A). Physiological characterization of the expulsion process in *E. coli* and species of *Streptococcus* has led to the conclusion that the process is dependent on some form of chemical energy, probably ATP (Chapter 6, Section B). This observation together with the recognition of an essential role of glycolytic intermediates led to the postulate that an ATP-dependent protein kinase-catalyzed reaction promoted sugar expulsion and that glycolytic intermediates stimulate efflux. This postulate led to the search for protein phosphorylation. In *Streptococcus pyogenes*, a single protein was shown to be phosphorylated under the conditions which specifically promoted sugar efflux. That protein was identified as HPr of the PTS, and a seryl residue was shown to be derivatized (Chapter 6, Section C). The identification of HPr(ser)phosphate led to the possibility that carbohydrate transport, both uptake and efflux, may be regulated by protein phosphorylation in processes in which intracellular metabolites control the activities of the kinases. Such a possibility is depicted in Fig. 1.6. It is proposed that phosphorylation of a seryl residue in HPr inhibits the activity of the phosphotransferase system. Phosphorylation of this seryl residue in HPr is controlled by a metabolite-activated protein kinase and an HPr(ser)P phosphatase. By a related process, the same kinase, or a different protein kinase, may regulate the hydrolysis of cytoplasmic sugar phosphates and the subsequent expulsion of inducers via the Enzymes II of the PTS

(Fig. 1.6). The details of these processes will be discussed in Chapter 6.

The involvement of a protein kinase in the regulation of carbohydrate transport is reminiscent of the recent studies of Wang and Koshland demonstrating that *Salmonella typhimurium* contains a multiplicity of protein kinases and protein phosphate phosphatases, each with a restricted number of target protein substrates (Chapter 6, Section D). One of these targets proved to be isocitrate dehydrogenase, a key enzyme whose activity regulates the relative activities of the tricarboxylic acid cycle and the glyoxylate shunt (Chapter 6,

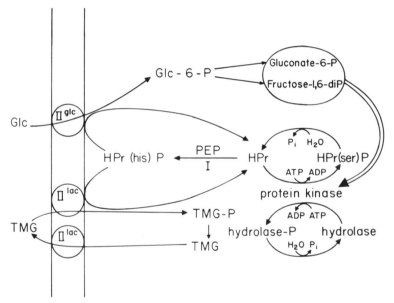

Fig. 1.6 Proposed involvement of ATP-dependent protein kinase in the regulation of carbohydrate uptake (inducer exclusion) and sugar efflux (inducer expulsion) in gram-positive bacteria such as the streptococci. The protein kinase phosphorylates a seryl residue in HPr, the small phosphoryl carrier protein of the phosphotransferase system. Phosphoseryl-HPr is a less efficient phosphoryl carrier protein than is the nonderivatized protein because it is poorly phosphorylated by Enzyme I. A protein kinase-catalyzed phosphorylation event is also thought to activate a sugar-phosphate hydrolase, which causes the hydrolysis of cytoplasmic sugar phosphates. The scheme illustrates the regulation of uptake of thio β-methylgalactoside (TMG) via the PTS as well as the hydrolysis of TMG-6-P and its expulsion from the cell. Because the kinase is activated by metabolic intermediates such as fructose 1,6-diphosphate and 6-phosphogluconate, the uptake and metabolism of another carbon source, such as glucose, [mediated by the glucose Enzyme II (II^{glc})] can abolish TMG accumulation mediated by the lactose Enzyme II (II^{lac}).

Section E). Further, some evidence suggests that protein kinase-catalyzed phosphorylation of phosphofructokinase may regulate the activity of this key glycolytic enzyme in *Bacillus subtilis*. These observations lead to the possibility that while cyclic AMP and inducer levels regulate the rates of transcription of carbohydrate transport and catabolic enzyme systems, metabolite-sensitive protein kinases may function as general regulators of the activities of these systems. In view of the central role of cyclic AMP and protein kinases in regulating metabolic and differentiative functions in eukaryotic cells, these observations may be of great evolutionary significance (Chapter 6, Section F).

Transcriptional regulatory mechanisms are as diverse as the transport proteins synthesized. Recent work has implicated transmembrane receptor and transport proteins in the control of transcriptional initiation. These processes will be discussed in Chapter 7. It will be seen that the involvement of integral membrane proteins allows extracellular substrates to function as transcriptional regulatory agents. Several distinct mechanisms have been proposed in order to account for exogeneous induction, and preliminary evidence suggests that several well-characterized transport proteins may play a direct role.

In order to understand the transport processes and the processes regulating carbohydrate uptake, salient features of the permease mechanisms must be understood. These features will be discussed in Chapters 2 and 3. The most significant structural and mechanistic advances made within recent years will be emphasized.

2

Mechanisms of Carbohydrate Transport

As discussed in a recent review (Dills *et al.*, 1980) carbohydrates are transported into bacterial cells by five distinct mechanisms (Fig. 2.1). Glycerol crosses the cytoplasmic membrane by diffusion through a nonstereospecific proteinaceous pore (Fig. 2.1,A). The lactose and melibiose permeases (Figs. 2.1,B and C) function by H^+ and Na^+ symport, respectively, while maltose is accumulated in *Escherichia coli* by an active transport system consisting of four distinct proteins (Fig. 2.1,D). These proteins function as a unit to drive the uptake of maltose and its higher homologues. Finally, mannitol is transported by the phosphotransferase system, consisting of two general energy-coupling proteins, Enzyme I and HPr, and a sugar-specific transport protein, the mannitol-specific Enzyme II.

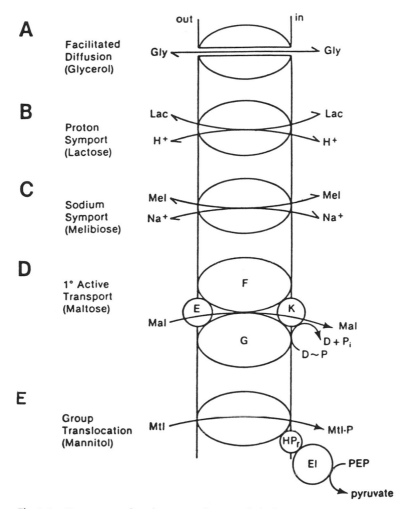

Fig. 2.1. Permease-mediated transmembrane carbohydrate transport processes that occur in *E. coli*. Glycerol (Gly) crosses the membrane by passive diffusion through a nonstereospecific protein pore. Lactose (Lac) and melibiose (Mel) enter the cell by cation symport involving protons and sodium ions, respectively. Each of these three transport processes is probably catalyzed by a single protein species encoded by a specific gene on the bacterial chromosome. Maltose (Mal) transport may involve the hydrolysis of a high energy phosphoryl donor (D ~ P), which drives the active accumulation of the sugar. Four distinct proteins, encoded by four different genes within the *mal* regulon, are required for transport. Finally, the transport of mannitol (Mtl) involves three proteins of the phosphotransferase system, the general energy coupling proteins of the system, Enzyme I (EI) and HPr, as well as the integral Enzyme II specific for mannitol (II^{mtl}). Each of these proteins is sequentially phosphorylated at the expense of phosphoenolpyruvate (PEP) before translocation of mannitol and its coupled phosphorylation can occur. [Reproduced with permission from Saier and Jacobson, (1984).]

Enzyme II^{mtl} catalyzes the concomitant transport and phosphorylation of its sugar substrate (Fig. 2.1,E). These five systems, each of which is the best characterized representative of its class of carbohydrate permeases, will be discussed in this and the subsequent chapter in order to acquaint the reader with the most significant recent developments dealing with the mechanisms of these transport processes and the structures of the transport proteins.

A. GLYCEROL PERMEASE: FACILITATED DIFFUSION

In virtually all prokaryotes (Lin, 1976) and eukaryotes (Lin, 1977) which have been studied to date, exogenous glycerol enters the cytoplasm by passive or facilitative diffusion. Two different mechanisms of facilitation appear to operate in eukaryotes and prokaryotes, respectively. Eukaryotic cells, such as human erythrocytes, transport glycerol by a saturable process, possibly by a carrier-type mechanism in which the substrate binding site of the permease is capable of movement, so that it can alternate between one state accessible to the internal side of the membrane and another state accessible to the cytoplasmic side of the membrane (Lin, 1977). By contrast, glycerol crosses the *E. coli* cytoplasmic membrane by a nonsaturable proteinaceous pore-type mechanism (Lin, 1976; Heller *et al.*, 1980). The glycerol facilitator is encoded by the *glpF* gene which maps at approximately 87 min on the *E. coli* chromosome (Berman-Kurtz *et al.*, 1971). This gene and the *glpK* structural gene, encoding glycerokinase, together with an operator-promoter region apparently comprise a single operon which is under the control of the glycerol repressor, the *glpR* gene product (Cozzarelli *et al.*, 1968; Hayashi and Lin, 1965).

Heller *et al.* conducted detailed kinetic studies defining the substrate specificity and transport characteristics of the glycerol facilitator (Heller *et al.*, 1980). An osmotic method and radioisotope techniques were employed to show that the *glpF* gene product allowed the entry of straight chain polyols including tetritols, pentitols and hexitols in addition to glycerol. The rates of transport were related primarily to the size and shape of the substrate and not to the stereo arrangement of the hydroxyl groups. For example, pentitols were transported more rapidly than hexitols. Further, while D- and L-arabinitol were transported at comparable rates, the other pentitols tested, ribitol and xylitol, were transported substantially more rapidly. Of the hexitols tested, galactitol and glucitol were transported about threefold more rapidly than mannitol. Ring sugars did not cross the membrane, but urea, glycine and glyceraldehyde, all straight-chain compounds of differing properties, were transported. Phosphorylated compounds could not permeate the membrane via the facilitator.

Other properties of the system similarly indicated that a fairly nonspecific pore-type mechanism was operative:

1. The system catalyzed both uptake and efflux of [^{14}C]xylitol at comparable rates.

2. [^{14}C]Xylitol uptake was not inhibited by a 100-fold excess of glycerol (500 mM).

3. A 1000-fold excess of xylitol (20 mM) decreased the rate of [^{14}C]glycerol uptake by only 40%.

4. The rate of [^{14}C]xylitol entry via the glycerol facilitator was practically unaffected by a 23°C drop in temperature.

5. The turnover number of the facilitator was about 2×10^5 molecules per second per protein molecule, a rate which is about 3 orders of magnitude greater than the corresponding value for the lactose permease. The latter permease probably functions by a carrier-type mechanism (see Section B).

All of these observations lead to the conclusion that the glycerol facilitator exists as a transmembrane channel with an inner diameter of about 0.4 nm (calculated from molecular radii of the substrates). Although channels with low transport specificity have been characterized in the outer membranes of gram-negative bacteria, (Table 1.1), the glycerol facilitator is the only known example of such a protein in the cytoplasmic membrane. Its specificity for unphosphorylated, neutral, straight-chain compounds explains the retention of normal intracellular metabolites when expression of the *glpF* protein is induced to high levels.

B. LACTOSE PERMEASE: SUGAR:H$^+$ SYMPORT

The lactose (*lac*) permease is the best understood transport system that functions by substrate–cation symport (Kaback, 1983; Hengge and Boos, 1983). As described in an earlier article (Dills *et al.*, 1980), a chemiosmotic energy coupling mechanism and a transport stoichiometry of lactose to proton of 1 : 1 as originally proposed by Mitchell (1963; 1967) appears to be well established. The coupling of lactose to H^+ movements is essentially obligatory so that transmembrane translocation of one substrate without the other does not occur under normal conditions (Dills *et al.*, 1980; Garcia *et al.*, 1982; Kaczorowski and Kaback, 1979; Page and West, 1980, 1981; Wright *et al.*, 1981). Thus, the substrate combining site of the free carrier (C) or the ternary complex of the carrier with the sugar and proton substrates (CSH^+) can move across the membrane, but the two possible binary complexes (CS or CH^+) are not transported at an appreciable rate. Only under conditions of high internal lactose concentration has lactose exit via the permease been reported to occur without a proton (Booth *et al.*, 1981; Ahmed and Booth, 1981). An alternative mechanism, in which translocation of the sugar and proton are temporally distinct, has been proposed (Lancaster, 1982).

In distinguishing a "carrier"- from a "channel"-type transport mechanism (Dills *et al.*, 1980; Leonard *et al.*, 1981), it is important to establish the number of substrate binding sites. Thus, a carrier-type mechanism would involve a single sugar binding site and a distinct proton binding site, both of which could assume coupled but alternative orientations with nonsimultaneous exposure to the external and internal faces of the membrane (Fig. 1.1). A channel-type mechanism might possess zero, one, two, or more substrate binding sites, depending on the mechanism, as discussed previously (Fig. 1.1; Leonard *et al.*, 1981). In some cases, such as proton translocation through bacteriorhodopsin (for reviews, see Leonard *et al.*, 1981; Ovchinnikov, 1982), there may be more than two substrate binding sites. The absence of a substrate binding site would result in a nonspecific pore such as exists for the glycerol permease. One solute binding site with *simultaneous* exposure to the internal and external faces of the membrane could give rise to a stereospecific pore. Two binding sites, one exposed to each side of the membrane, would provide a channel-type shuttle mechanism for the translocation of solute. The succinate permease in *E. coli* and other binding protein-dependent systems may be examples of this last type of mechanism (Leonard *et al.*, 1981; Lo, 1977).

Detailed kinetic studies of Kaczorowski *et al.* (1979; Kaczorowski and Kaback, 1979) showed that while unidirectional lactose efflux from *E. coli* membrane vesicles was highly dependent on the medium pH, bidirectional counterflow with saturating concentrations of lactose was much more rapid and was independent of external pH. Only the unidirectional flux of lactose was accompanied by the transient formation of a membrane potential (negative inside). The results were interpreted in terms of a carrier-type mechanism as previously defined. In the unidirectional efflux process, lactose and a proton bind to their respective sites exposed to the cytoplasmic surface of the membrane and are translocated as a ternary complex such that the complex becomes exposed to the external face of the membrane. Lactose, followed by the proton, then dissociates into the external medium, and the free carrier can cycle back to the inner face of the membrane. Proton release, or translocation of the free carrier, is presumed to be rate limiting for the overall process. In counterflow, deprotonation need not occur; the protonated carrier cycles with rapid dissociation of the lactose–carrier complex at both the cytoplasmic and external surfaces.

More recently lactose transport has been shown to exhibit similar kinetic properties employing proteoliposomes in which the purified lactose carrier protein was reconstituted in unilamellar phospholipid vesicles (Garcia *et al.*, 1983). The reconstitution procedure employed involved, first, octyl glucoside dilution (Newman and Wilson, 1980; Newman *et al.*, 1981), second, freezing and thawing, and finally, sonication (Garcia *et al.*, 1983). The principal observations made with *E. coli* membrane vesicles (Kaczorowski and Kaback, 1979) were reproduced. Thus, unidirectional lactose efflux was strongly pH dependent, showing a 20-fold increase as the pH was raised from 5.5 to 7.5, although lactose exchange (counterflow) was pH independent and much more rapid. As expected, only unidirectional transport resulted in proton symport and the transient generation of a membrane potential. These observations, as well as marked deuterium solvent isotope effects [observed for unidirectional lactose efflux, but not for counterflow or membrane potential driven lactose uptake (Viitanen *et al.*, 1983)] were easily interpretable in terms of the carrier-mediated, preferentially ordered mechanism of lactose–H⁺ symport proposed by Kaczorowski *et al.* (1979; Kaczorowski and Kaback, 1979). The fact that similar turnover numbers and kinetic results were obtained with *E. coli* membrane vesicles and with proteoliposomes reconstituted with the purified lactose carrier attests to the fact that this protein alone is responsible for the translocation of lactose across the cytoplasmic membrane of

E. coli. A similar model has been suggested by Overath and Wright (1983).

Transport and direct binding studies, conducted by Wright *et al.* (1981) with an *E. coli* strain possessing greatly enhanced levels of the lactose permease, led to similar conclusions. Moreover, it was shown that under the conditions employed, the equilibrium constant for galactoside binding was independent of substrate proton binding, and galactoside binding did not elicit proton binding. These observations suggest that formation of the ternary complex (CSH^+) can occur by a random, nonordered mechanism and that the two substrates (galactoside and H^+) bind independently (noncooperatively) of each other. It should be noted, however, that the experiments of Wright *et al.* were conducted at pH 7.5, and since the pK_a of the residue responsible for transporting the proton may be near 8.3 (Kaczorowski *et al.,* 1979), it appears that conditions were not optimized for measuring an effect of galactoside binding on the carrier–proton association process. The conclusion of a random, nonordered mechanism is not necessarily in conflict with the suggestion of Kaczorowski *et al.* that during efflux of sugar, lactose dissociation from the ternary complex *preferentially* precedes proton dissociation. An equilibrium situation was studied by Wright *et al.,* whereas Kaczorowski *et al.* examined a kinetic situation. It is worth noting, however, that Page and West (1980, 1981) conducted kinetic studies of uptake in whole cells at various pH values and in the presence of various competing galactosides. While the data were obtained under energized conditions, and the interpretation of their results is complex, a random mechanism of galactoside and proton binding was suggested.

Of further significance, the studies of Wright *et al.* established that in contrast to some earlier suggestions (reviewed in Dills *et al.,* 1980), the permease possesses a single galactoside binding site. Three independent results led to the following conclusions:

1.   *p*-Nitrophenyl α-galactoside [a class II substrate (see Dills *et al.,* 1980)] and 1-thio-β, β-digalactoside (a class I substrate) bound to the carrier with similar stoichiometries, probably one sugar molecule per polypeptide chain (Overath *et al.,* 1979; Teather *et al.,* 1980).

2.   All substrates protected the essential sulfhydryl group(s) of the permease at concentrations which were consistent with their relative binding affinities.

3. Each galactoside studied competed for binding with the other binding substrates, and binding of and inhibition by each sugar was characterized by a single dissociation constant. It was suggested from these and other studies (see Dills *et al.*, 1980) that the carrier is approximately functionally symmetrical with respect to substrate binding, at least in the absence of an imposed membrane potential. In this regard, it is interesting to note that the action of proteolytic enzymes on right-side-out and inside-out membrane vesicles inactivates *lac* carrier function in an apparently symmetrical fashion (Goldkorn *et al.*, 1983). However, employing a kinetic approach with membrane vesicles, Kaczorowski *et al.* (1979) showed that under de-energized conditions, the kinetic constants for lactose transport differed for the influx and efflux processes. This result suggests functional as well as structural asymmetry.

The studies described above, particularly those of Wright *et al.* (1981), provided detailed information concerning the galactoside binding site, but not the proton site. Relevant to the nature of the proton binding site, Garcia *et al.* found that reagents which inactivate histidyl residues in proteins (diethylpyrocarbonate or light plus rose bengal) inactivated active lactose uptake as well as counterflow. Galactoside substrates of the *lac* permease prevented this inactivation. Acylation of a single histidyl residue with diethylpyrocarbonate appeared to be sufficient to inactivate the carrier. Surprisingly, neither reagent inhibited either binding of the sugar substrate to the carrier or facilitated diffusion. In fact, facilitated diffusion of lactose was stimulated (Patel *et al.*, 1982) in a fashion reminiscent of the energy uncoupled *lac* Y mutants of Wilson and co-workers (1970; Wilson and Kusch, 1972). Further, the reagent blocked lactose-induced proton influx, suggesting that acylation of a histidyl residue altered the proton binding site, obviating a need for sugar:proton symport. These results lead to the suggestion that a histidyl residue in the lactose carrier may function directly in transmembrane proton translocation, possibly as the proton binding site. If this histidyl residue is truly the binding site for the translocated proton, it should be accessible both from the inside and the outside of the cell. Unfortunately, membrane-impermeable derivatives of diethylpyrocarbonate are not yet available to test this hypothesis.

As a consequence of the recent studies summarized above, a sugar:proton symport mechanism, as originally proposed by Mitchell, or a single-site exposure (carrier) model (Lancaster, 1982) rather than a gated pore model (Lombardi, 1981) appears to be favored. It has been suggested that the proton electrochemical gradient alters the distribution of the lactose carrier between two different kinetic states (Robertson *et al.*, 1980). One state, typified by the energized carrier when it catalyzes active transport, possesses high affinity for its galactoside substrates. The other state, typified by the de-energized carrier when it catalyzes facilitated diffusion, possesses low affinity for these sugars. How energy functions to interconvert the two permease states is not at all clear. One possibility would be an interaction of the electric field with the dipole moment of the permease, inducing a conformational change with altered maximal activity or affinity for the substrate.

A possible consequence of a membrane potential-induced conformational change is a change in oligomeric structure of the permease. Such an alteration would be predicted if the potential-induced protomeric conformation of the *lac* permease possesses higher affinity for another protomer than the relaxed form of the protein. Precisely these consequences of an imposed proton electrochemical gradient have recently been postulated (Goldkorn *et al.*, 1984). *E. coli* cytoplasmic membrane vesicles were irradiated with a high energy electron beam at $-135°C$, and the time-dependent inactivation of the lactose permease was studied as a function of radiation dosage. Application of target theory allowed determination of a functional molecular mass. When nonenergized vesicles were irradiated, the functional molecular weight of the permease was 45 kilodaltons in agreement with the known molecular weight of the *lacY* gene product. By contrast, when the same procedure was performed with vesicles that were energized prior to freezing and irradiation, a functional molecular weight of about 90 kilodaltons was observed. The results suggested that imposition of a proton electrochemical gradient induces dimerization of the normally monomeric *lac* carrier.

The possibility that the proton electrochemical gradient induces dimerization is not favored by all investigators (Wright *et al.*, 1983). The protein appeared to be a monomer in detergent micelles of dodecyl-maltoside as determined by sedimentation equilibrium in the analytical ultracentrifuge. Further, an earlier suggestion that certain mutants defective in the *lacY* gene product exhibited a negative-dominant phenotype (Mieschendahl *et al.*, 1981) could not be confirmed (Wright *et al.*, 1983). These results, however, neither support nor disprove the proposal of Goldkorn *et al.* (previously suggested by Robertson *et al.*, 1980) that the electrochemical gradient promotes dimerization of the *lac* carrier.

It has been suggested that one function of chemiosmotic energy (in this case, the proton electrochemical gradient) might be to alter the affinity of the *lac* carrier for its sugar substrates. Konings and Robillard (1982) have recently re-examined this possibility in light of new data obtained while studying the effects of redox state of sulfhydryl groups on transport activity. Lipophilic oxidizing agents such as plumbagin were found to inhibit lactose transport, and this inhibitory effect was reversed by addition of thiols such as dithiothreitol. The same lipophilic oxidizing agents prevented irreversible inactivation of the carrier by *N*-ethylmaleimide. The dithiol-specific reagent, phenylarsine oxide, was also inhibitory at low concentrations. Based on these results, the redox-sensitive step was suggested to be the conversion of a dithiol to a disulfide.

A kinetic analysis of ascorbate/phenazine methosulfate-driven lactose uptake in membrane vesicles revealed that plumbagin increased the K_m value from 0.2 to 20 mM while decreasing the V_{max} to about one-half of the original value. Worthy of note was the fact that de-energization of the vesicles (dissipation of the proton electrochemical gradient) yielded a similar hundred-fold shift in apparent K_m, from 0.2 to 20 mM. It was suggested that the proton electrochemical gradient, or one of its components, alters the ligand affinities of the carrier during a single transport cycle by converting the carrier from the oxidized to the reduced form. In other words, the authors suggested that reversible oxidation–reduction of sulfhydryl groups in the carrier occurs as part of the normal cycle of solute–proton symport (Konings and Robillard, 1982). The cycle was envisaged as follows:

1. Substrate binds to the outer site initially in the high-affinity form.
2. Binding of H$^+$ causes the carrier to switch to the oxidized form due to a shift in the apparent pK of the redox couple.
3. Hydrogen moves down the hydrogen-conducting pathway, reducing the inner site, and the ligand is drawn to the inner high-affinity site.

4. Proton and solute are released into the medium when the inner site is oxidized, and the electron is carried back to the outer site by an electron conducting pathway.

The principal postulate summarized above suggests that an oxidation–reduction cycle occurs as an essential part of the active transport process. Facilitated diffusion, which involves only low affinity binding of lactose to the carrier should be unaffected by the lipophilic oxidizing agents that block active uptake. Moreover, treatment of the carrier with diethylpyrocarbonate, which abolishes lactose: proton symport while still permitting lactose facilitated diffusion (Garcia *et al.*, 1982; Patel *et al.*, 1982), should have the same effect as the oxidizing agents. It is possible, however, that the oxidized disulfide form of the carrier is inactive while only the reduced (dithiol) form possesses activity and exhibits substrate binding characteristics. If this alternative interpretation is correct, sugar substrates of the permease may influence the equilibrium between the oxidized and reduced forms, favoring the reduced form. This shift in equilibrium at high substrate concentrations may be responsible for the apparent high K_m activity in the presence of oxidizing agents. If this postulate is correct, oxidizing agents such as plumbagin would be expected to inhibit facilitated diffusion in the absence of energy or in the presence of diethylpyrocarbonate. Further, the redox events observed by Konings and Robillard may function solely to regulate the activity of the system, rather than as an integral component of the cyclic catalytic transport mechanism (see Chapter 4, Section A). This possibility had, in fact, been suggested earlier from studies which showed that the transmembrane electrical potential influences the reactivity or accessibility of essential sulfhydryl groups within the *lac* carrier to irreversible derivatization by *N*-ethylmaleimide (Cohn *et al.*, 1981). Further studies will be required to distinguish these possibilities.

The kinetic and mechanistic studies reported above provide little information concerning the structure of the protein or protein complex which catalyzes lactose–proton symport. However, the *lacY* gene, encoding the lactose permease, has been cloned (Teather *et al.*, 1980), and the nucleotide sequence of the gene has been determined (Büchel *et al.*, 1980).

Expression of the cloned *lacY* gene in an *in vitro* protein synthesizing system gave rise to a protein which appeared identical to the lactose permease isolated from the cytoplasmic membrane with respect to its molecular weight and its N-terminal amino acid sequence (Ehring *et al.*, 1980). The amino acid composition of the purified permease was in fair agreement with that predicted from the nucleotide sequence, and the N-terminal amino acid sequence was in agreement with that predicted from the *lacY* nucleotide sequence assuming that the N-terminus was not processed. Using carboxypeptidase, the purified protein has also been shown to possess the carboxyl terminal aminoacyl sequence predicted from the nucleotide sequence (T. H. Wilson, personal communication). This terminal region can be removed by treatment with carboxypeptidase without loss of activity (N. Carrasco, R. Mitchell, D. Herzlinger, and H. R. Kaback, unpublished results). Based on the DNA sequence of the *lacY* gene and studies with the purified protein, the permease apparently consists of a single polypeptide chain of 417 amino acid residues (71% nonpolar) having a molecular weight of 46,504. It is not processed at either the N- or the C-terminus.

The amino acid sequence of the protein has been analyzed for sequential hydropathic character by the computer method of Kyte and Doolittle (Foster *et al.*, 1983; Kyte and Doolittle, 1982). The analysis revealed that the protein consists of 12 extended hydrophobic segments with a mean length of 24 ± 4 residues per segment (Fig. 2.2). Both the C- and N-termini appear to be localized to the cytoplasmic surface of the membrane (Seckler *et al.*, 1983; K. Beyreuther, unpublished results). Based on circular dichroic measurements conducted with the purified lactose permease protein, approximately 85% of the amino acid residues are probably arranged in helical secondary structure (Foster *et al.*, 1983). Since about 70% of the 417 amino acid residues were found to be included within the 12 hydrophobic segments, it was predicted that the imbedded segments are largely α-helical, with each helix extending the entire thickness of the membrane, perpendicular to the plane of the membrane. This structure is reminiscent of the seven-helix structure previously described for bacteriorhodopsin (Henderson and Unwin, 1975; Henderson, 1975; Long *et al.*, 1977) and the seven-helix structure proposed for the mannitol Enzyme II of the phosphotransferase system (see Chapter 3, Section E).

Biochemical studies have provided initial substantiation for the general structural model proposed above. Thus, monoclonal antibodies directed against the lactose carrier protein have been isolated

LACTOSE CARRIER PROTEIN

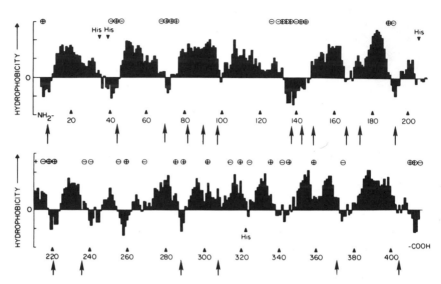

Fig. 2.2. Hydropathic profile of the *lac* carrier protein. A modified version of the hydropathy program described by Kyte and Doolittle (1982) was used. The program assigns values to amino acids on the basis of their hydropathic character. Using a moving-segment approach that continuously determines the average hydropathy of a segment of seven residues as it advances through the sequence, consecutive scores are plotted at the middle of each segment from the amino to the C-terminus. Regions predicted to form reverse turns by secondary structure predictions of Chou and Fasman (1978) are designated by arrows. The positions of all histidines and charged residues are noted. [From Foster *et al.* (1983), with permission.]

which are directed against epitopes (antibody combining sites), which are accessible only to the external surface of the membrane (Carrasco *et al.*, 1982). The protein must therefore be asymmetrically inserted in the membrane in a highly specific fashion. Moreover, proteolysis of sealed *E. coli* vesicles of either a right-side-out or inside-out orientation was shown to inactivate the carrier (Goldkorn *et al.*, 1983). Surprisingly, inactivation from the two sides of the membrane occurred with similar kinetics, and the electrophoretic mobilities of the cleaved fragments were similar, regardless of vesicle orientation. Detergent solubilization of the protein followed by protease digestion resulted in much more extensive degradation to smaller peptides. Proteolysis in intact membranes, which abolished transport function, did not interfere with substrate binding as indicated by galactoside photoaffinity-labeling experiments (Goldkorn *et al.*, 1983). These results clearly show that the *lac* carrier spans the membrane and suggest that the galactoside binding site resides within a segment of the protein that is embedded within the phospholipid bilayer.

The above discussion tacitly assumes that the single polypeptide chain encoded by the *lacY* gene is both necessary and sufficient for catalysis of lactose transport across the cytoplasmic membrane of *E. coli*. This fact has been established by reconstitution of the purified protein in proteoliposome vesicles consisting solely of the lactose permease and phospholipids (Newman and Wilson, 1980; Newman *et al.*, 1981). Solubilization and reconstitution were effected by a modification of the octylglucoside dilution procedure of Racker *et al.* (1979). The protein was solubilized from the membrane with octylglucoside (1.25%) in neutral phosphate buffer containing dithiothreitol, lactose, and *E. coli* phospholipids to stabilize the protein. Subsequent centrifugation at 175,000 g for 1 hr removed residual particulate materials while leaving the permease protein and 50% of the total membrane protein in solution. Transport activity could be reconstituted in presonicated liposomes by sonication in the presence of 1.25% octylglucoside followed by a 30-fold dilution into detergent-free buffer. After centrifugation at 85,000 g for 1.5 hr the resuspended vesicles exhibited both membrane potential-driven lactose uptake and more rapid lactose counterflow. About 10% of the original membrane protein became associated with the liposomes.

Lactose transport in the reconstituted liposomes exhibited the expected properties based on the known characteristics of the lactose permease: facilitation of diffusion in the absence of energy, active uptake with an imposed membrane potential (negative inside), and rapid and transient counterflow when the proteoliposomes were preloaded with nonradioactive lactose. Thiodigalactoside was a potent competitive inhibitor of lactose uptake into proteoliposomes, and transport was sensitive to inhibition by *N*-ethylmaleimide, an irreversible sulfhydryl reagent. No activity was observed when membranes from a *lac Y⁻* strain were used for solubilization of the carrier. Reconstitution and a rapid filtration assay therefore provided a reliable method for following purification of the active permease.

Purification of the lactose permease was facilitated by the use of a bacterial strain which overproduced the protein and the availability of a photoaffinity reagent, 4-nitro[2-^{3}H]phenyl α-D-galactopyranoside (Kaczorowski et al., 1980). The radioactively labeled protein, inactivated with this reagent, co-purified with the functionally active permease. The purification procedure was as follows: other proteins were first removed from the E. coli membranes by sequential extraction with 5 M urea followed by 6% cholate. Subsequent extraction with octylglucoside gave an active, solubilized fraction with 12-fold higher specific activity than the original membrane preparation. A single DEAE Sepharose ion exchange column chromatographic step then yielded a protein which was electrophoretically pure. A 35-fold purification was attained. Although the protein has a molecular weight of 45,000, based on the sequence analysis, its migration rate during sodium dodecyl sulfate polyacrylamide gel electrophoresis was that of a 33,000-dalton protein. This abnormally high migration rate had been noted previously for the lactose carrier as well as for certain other integral membrane permeases (see Chapter 3, Section E). It presumably reflects enhanced binding of the detergent to the hydrophobic side chains of the protein or in complete unfolding of the polypeptide chain.

The amino acid composition of the purified lactose carrier was close to that predicted from the lacY gene sequence. An exception was the low value reported for isoleucine (24 as compared with 32 from sequence analysis). Low values for the branched chain amino acids have been repeatedly observed for integral membrane proteins (R. A. Nicholas, personal communication; see Lee and Saier, 1983b, for a discussion of this phenomenon).

The highly purified protein reconstituted in proteoliposomes exhibited all of the vectorial transport properties expected of the lactose carrier (i.e., membrane potential driven lactose accumulation, counterflow, lactose-dependent alkalinization of the medium, inhibition by specific inhibitors). Significantly, the turnover numbers of the carrier as well as the substrate K_m values were similar for all three processes (active uptake, counterflow, and facilitated diffusion) when reconstituted proteoliposomes were compared with bacterial membrane vesicles (Kaback, 1983). Additionally, proteoliposomes simultaneously reconstituted with the lac permease and purified O-type cytochrome oxidase catalyzed electron transfer-driven active lactose accumulation (Matsushita et al., 1983). Consequently, it can be concluded that a single polypeptide species, the product of the *lacY* gene, is alone reponsible for all of the catalytic functions attributed to the lactose permease including active uptake in response to a proton electrochemical gradient. The possibility of an involvement of additional components as discussed by Hengge and Boos (1983) can be discounted.

Finally, it is interesting to note that the natural substrate for which the lactose catabolic enzyme system (permease plus β-galactosidase) evolved may have been glycerol-β-galactoside instead of lactose [see Hengge and Boos (1983) for a discussion of this possibility]. The former galactoside is released in large quantities as a result of lipase action on plant galactolipids in the intestines of plant-eating animals. These galactolipids are far more widespread in nature that is lactose which is found almost exclusively in mammalian milk. In support of this notion is the observation that while lactose itself is not an inducer of the *lac* operon, glycerol-β-galactoside is a potent inducer, even at micromolar concentrations. Moreover, virtually all *lac* permeases studied in a variety of gram-positive and gram-negative bacterial strains exhibit broad substrate specificity. If this line of reasoning is truly reflective of the evolutionary pressures which gave rise to the lactose catabolic enzyme system, the more general designation "β-galactoside catabolic system" would be appropriate.

C. MELIBIOSE PERMEASE: SUGAR : Na⁺ SYMPORT

Early experiments established that the melibiose (*mel*) permease in *Salmonella typhimurium* can couple substrate uptake to Na⁺ uptake with a stoichiometry of $1:1$, and that sugar–cation symport can account for energization of the system (for a review, see Dills *et al.*, 1980). Some evidence suggests that the lactose and melibiose permeases exhibit common structural and functional attributes and therefore are evolutionarily related. The evidence is as follows:

1.   Both permeases function by cation co-transport, probably by a similar carrier-mediated mechanism (Dills *et al.*, 1980).

2.   While the lactose permease functions exclusively by sugar–H⁺ co-transport, the melibiose permease couples sugar uptake to the symport of H⁺, Li⁺, or Na⁺, depending of the conditions, the bacterial strain, and the sugar substrate under study (Tsuchiya and Wilson, 1978; Tsuchiya *et al.*, 1978, 1983).

3.   Single point mutations within the *mel* operon (probably within

the structural gene for the melibiose permease) can change the cation specificity of the carrier (Niiya *et al.*, 1982).

4.   Both transport systems exhibit broad and overlapping sugar substrate specificities (Dills *et al.*, 1980).

5.   The two systems can be solubilized and reconstituted in a proteoliposome system by essentially the same procedure (Tsuchiya *et al.*, 1982). This fact implies that they exhibit structural similarities.

6.   Genetic studies lead to the probability that a single structural gene encodes the melibiose permease (Schmitt, 1968; Levinthal, 1971), and the same fact is well established for the lactose permease (see Section B).

7.   Both systems are regulated by the phosphoenolpyruvate-dependent phosphotransferase system by a similar mechanism involving cooperative substrate–Enzyme III[glc] binding to the permease (Dills *et al.*, 1980; Saier, 1982; Saier *et al.*, 1983; see Chapter 4, Section C).

In view of these observations, this author questions the proposal of Zilberstein *et al.* (1982) that the melibiose permease and other Na$^+$ symport permeases in bacteria possess a dimeric structure, one subunit of which is common to all Na$^+$ symporters and catalyzes Na$^+$ translocation. The successful solublization and reconstitution of melibiose uptake activity in proteoliposomes (Tsuchiya *et al.*, 1982) should allow purification of the protein and establishment of the translocation and energy coupling mechanisms.

DNA sequencing of the cloned structural gene of the melibiose permease (Tsuchiya *et al.*, 1982) has recently been completed (T. Tsuchiya, unpublished results). While the two permeases show little sequence homology, the hydropathy patterns show striking similarities suggesting a common origin. Tertiary structure has frequently been found to be a reliable indicator of relatedness when sequence divergence is extensive. The possibility of convergent evolution cannot be ruled out, however.

In spite of the numerous similarities noted above for the *lac* and *mel* permeases, several differences have been noted (Cohn and Kaback, 1980):

1.   While the *lac* permease exhibits strict cation specificity, that of the *mel* permease is relatively broad.

2.   While lactose and H$^+$ binding to the *lac* carrier appear to

occur independently of one another (Wright *et al.*, 1981), Na⁺ and melibiose binding to the *mel* carrier is apparently cooperative (Tokuda and Kaback, 1978).

3. While unidirectional lactose transport is slow relative to exchange transport (counterflow) (Kaczorowski *et al.*, 1979), these two rates are comparable for the *mel* permease (Cohn and Kaback, 1980).

4. Rates of lactose efflux, but not exchange transport are altered by imposition of a proton electrochemical gradient (Kaczorowski *et al.*, 1979). By contrast, rates of melibiose exchange as well as efflux, catalyzed by the melibiose permease, are influenced similarly (Tokuda and Kaback, 1978).

5. Imposition of a proton electrochemical gradient or a lactose concentration gradient across the membranes of bacterial vesicles decreases the apparent K_m of the *lac* permease for lactose by two orders of magnitude. There is only a slight effect on the maximal velocity of transport. On the other hand, vesicle energization leads to a large increase in the V_{max} of sugar uptake via the *mel* permease with no change in the apparent K_m of the process (Cohn and Kaback, 1980).

6. Treatment of the *lac* carrier with the histidine-specific reagent, diethylpyrocarbonate, abolishes active lactose uptake and counterflow while stimulating facilitated diffusion of the sugar. The melibiose permease, on the other hand, responded to diethylpyrocarbonate treatment with a decrease in the V_{max} for energy-driven sugar uptake without affecting the K_m of the process.

These observations have led Cohn and Kaback (1980) to propose that the melibiose and lactose permeases differ in certain kinetic aspects, and possibly in certain fundamental mechanistic features. Specifically, they proposed that the unloaded *lac* carrier (C) bears a net negative charge while the loaded carrier (ternary complex, CSH⁺) is neutral. By contrast, the free *mel* permease may be neutral, whereas the ternary complex bears a net positive charge. It is noteworthy, however, that in both permeases, diethylpyrocarbonate interfered with the responses of the permeases to the relevant electrochemical ion gradient. Thus, the reagent prevents the energy-dependent decrease in the K_m of lactose entry via the *lac* carrier, while it interferes with the energy-promoted increase in the V_{max} of sugar uptake via the *mel* carrier.

Despite several interesting but possibly superficial differences, the two permeases may catalyze two mechanistically similar processes. Different substrates of the lactose permease exhibit markedly different kinetic behavior. For example, while lactose exhibits rates of counterflow which are greatly in excess of the unidirectional transport rate, the rate of thiodigalactoside exchange is comparable to the rate of unidirectional transport (Wright *et al.*, 1981). Explanations for these kinetic differences are not yet available, but they are presumed to represent differences in the relative values of kinetic constants rather than fundamental mechanistic differences.

D. MALTOSE PERMEASE: PRIMARY ACTIVE TRANSPORT

As a result of extensive genetic studies and some less detailed biochemical work, the *mal* regulon, encoding the maltodextrin permease proteins and the maltodextrin catabolic enzyme system, is fairly well defined (Bedouelle, 1984; Hofnung, 1982). The *malA* region at 74' on the *E. coli* chromosome consists of two operons: one, the *malT* operon on the right side, encoding a transcriptional activator protein; the other, on the left side, encoding the two maltose catabolic enzymes—amylomaltase (encoded by the *malQ* gene) and maltodextrin phosphorylase (encoded by the *malP* gene) as shown in Fig. 2.3. All of the proteins of the permease system are encoded within the *malB* region which consists of two operons, both of which are under the control of the *malT* protein as well as the cyclic AMP receptor protein (CRP) (Chapon, 1982; Guidi-Rontani *et al.*, 1982). It is interesting to note that these two operons have been reported to be expressed at different times in the cell cycle (Ohki *et al.*, 1982). One of the *malB* operons (right) consists of the *malK* and *lamB* (*malL*) genes, which respectively encode a constituent of the maltose permease in the cytoplasmic membrane (K) and the maltodextrin porin or lambda receptor in the outer membrane (L), hereafter designated maltoporin. Also included within this operon may be the small *molA* (*malM*) gene encoding a protein of unknown function. This author advocates changing the nomenclature of the (right) *malB* operon from the traditional designation of *malK lamB molA* to *malK malL malM*. The latter nomenclature will be used in subsequent sections of this book (see Chapter 3, Section B for a discus-

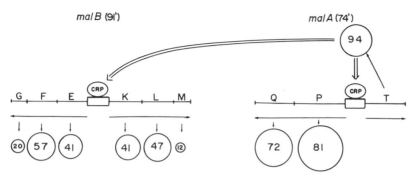

Fig. 2.3. Genetics (top) and biochemistry (bottom) of the maltose regulon, encoding the maltose catabolic enzyme system. The *malB* region, at 91 min on the *E. coli* chromosome, consists of two operons encoding the proteins which comprise the maltodextrin permease system. The *malE* gene encodes the periplasmic maltose-binding protein, which interacts with both the maltodextrin porin in the outer membrane (the lambda receptor, L) as well as the F–G pore complex found in the cytoplasmic membrane (E; MW = 41K); *malF* and *malG* encode two integral cytoplasmic membrane proteins, F and G, which together comprise the maltodextrin-specific channel in the cytoplasmic membrane. The molecular weight of the former protein is about 57K, and that of the latter protein is about 20K (H. Shuman, personal communication). *malK* encodes the peripheral cytoplasmic membrane protein (K; MW = 41K), which may bind to the cytoplasmic surface of the G-protein. The K-protein may function to energize the permease and to control the transport rate by interaction with the allosteric regulatory protein, the Enzyme III^{glc} of the phosphotransferase system (see Chapter 4). The gene, *malL* (or *lamB*) encodes the maltodextrin porin, which is also the receptor for phage lambda (L; MW = 47K). The presence of an additional gene, *malM* (or *molA*), encoding a 12K protein of unknown function, has been suggested from the nucleotide sequence of the *malB* region. (Note: the inclusion of three genes designated *malK*, *lamB*, and *molA* within a single operon is contrary to accepted genetic nomenclature. The preferred designations, employed in the figure are *malK*, *malL*, and *malM*, respectively (120,121).

The *malA* region, at 74 min on the *E. coli* chromosome, consists of two operons. The *malP malQ* operon encodes the enzymes, maltodextrin phosphorylase (P; MW = 81K) and amylomaltase (Q; MW = 72K). The *malT* gene encodes the specific transcriptional activator protein of the *mal* regulon (T; MW = 94K). Expression of all four operons in the *mal* regulon are under the control of the T protein and of the cyclic AMP receptor protein (CRP).

sion of preferred genetic nomenclature). The other operon, on the left, contains the three structural genes, *malE*, *malF* and *malG* which encode the periplasmic maltose-binding protein (E), as well as two integral cytoplasmic membrane constituents of the maltose permease, the F and G proteins. The approximate molecular weights of all of the proteins encoded by the *mal* regulon are presented in Fig.

2.3 while the presumed locations and orientations of these proteins in the bacterial cell are shown in Fig. 2.4. The amino acid sequences of several of these proteins have been deduced from the nucleotide sequences of the corresponding genes, and these sequences have been analyzed for their hydropathic character as described by Kyte and Doolittle (1982). These analyses are discussed in some detail below.

Of the five known proteins of the maltose permease system, only the lambda receptor, the outer membrane maltodextrin-specific porin (maltoporin), is reasonably well understood both structurally and functionally. Fine structure genetic analyses of mutations affecting the functions of the protein (Clement *et al.*, 1982) have been possible because of the availability of deletion mutations in the *lamB* (*malL*) gene. From the amino acid sequence of the protein (Clement *et al.*, 1982), sequential hydropathy analyses as well as comparisons

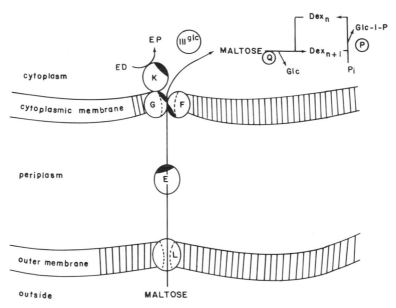

Fig. 2.4. Topology of the proteins of the maltodextrin permease and catabolic enzyme system in *E. coli*. Protein designations are in accordance with the genetic designations noted in Fig. 2.3. ED, energy donor(s); EP, energy product(s); Dex, maltodextrin; Glc, glucose; III^{glc}, Enzyme III^{glc} of the phosphotransferase system which binds to the *malK* protein to inhibit maltose uptake.

with other porin-type proteins in the outer membrane of *E. coli* have revealed a number of interesting structural features.

The precursor of maltoporin is a protein of 446 aminoacyl residues, 25 of which (the signal sequence) are cleaved from the N-terminus. The processed protein thus consists of 421 residues with 39 basic residues, 54 acidic residues, and no long, uncharged peptides. There are several possible alternative orientations of the protein in the outer membrane. First, it is possible that no single hydrophobic sequence traverses the membrane but that portions of the protein fold within the membrane in a complex way which involves many interchain interactions. Second, intramembranous α-helical portions of the protein may possess charged residues which function to line the hydrophilic channel and bind the substrate sugar residues. Relevant to this suggestion is the intriguing observation that charged residues frequently occur at intervals of 3–4, 7–8, and 11–12 residues along the chain. This observation could reflect an α-helical arrangement of part of the polypeptide within the membrane with charged, polar groups on one side of the helix, comprising part of the aqueous channel, with hydrophobic groupings on the outer side to interact with the hydrophobic matrix of the membrane. A similar arrangement has been postulated for the structure of parts of the human A-II high-density lipoprotein (see Saier and Stiles, 1975) as well as for Braun's lipoprotein in the outer membrane of *E. coli* (Lugtenberg and Van Alphen, 1983). Third, the most probable arrangement of the polypeptide backbone of maltoporin in the outer membrane of *E. coli* involves antiparallel β-pleated sheets instead of an α-helical array (Neuhaus, 1982; Garavito *et al.,* 1982). Such a structure has been demonstrated for the matrix porin in the outer membrane of *E. coli* (Rosenbusch, 1974), and circular dichroism measurements on maltoporin lead to the conclusion that the latter protein is similar in this regard (Neuhaus, 1982). Interchain hydrogen bonding among β-strands within the protein presumably neutralize the polarity of the peptide bonds within a hydrophobic environment. Relevant to this possibility is the observation that purified maltoporin, like the matrix porin, can be reconstituted with phospholipid and lipopolysaccharide to form an ordered, planar, hexagonal lattice (Yamada *et al.,* 1982; Neuhaus, 1982). Because the matrix porin exists as a trimeric structure and forms analogous hexagonal arrays, it is possible that native maltoporin is a trimer. In fact, studies of Nakai and Ishii (1982) and of Neuhaus (1982) aimed at measuring the molecular weights of maltoporin oligomers in detergent solutions reinforce this possibility. The recent crystallization of maltoporin from detergent solutions containing high salt (Garavito *et al.*, 1982) may allow a definitive solution to the problem of the tertiary and quaternary structure of the protein.

At a distance of about 300 bases downstream from the 3'end of the coding region of the *malL* (*lamB*) gene and on the same strand, is an unidentified reading frame of 131 codons (Clement *et al.*, 1982). This postulated structural gene has been designated *molA* (or *malM* in Fig. 2.3 according to accepted genetic nomenclature) and probably encodes a small periplasmic or outer membrane protein since the upstream portion of the nucleotide sequence encodes a typical signal peptide at the N-terminus of the postulated protein. The sequential hydropathy analysis reveals that this protein is fairly hydrophobic and therefore is probably associated with the outer membrane. In substantiation of this possibility, the *malM* gene was fused with *lacZ,* and the hypothesized hybrid protein was shown to be expressed (E. Gilson and M. Hofnung, unpublished results). It was localized to the envelope fraction of the cell, one-half being associated with the outer membrane. This result suggests that the native *malM* protein is probably associated with the outer membrane. Its function may therefore involve interaction with maltoporin.

The nucleotide sequence between *malL* and *malM* contains repetitive and palindromic sequences which if transcribed would give rise to stable hairpin structures which could influence expression of the *malL* and/or *malM* gene(s) (Clement *et al.*, 1982). Similar structures have been noted for the intercistronic regions of other operons (Higgins *et al.*, 1982; Lee and Saier, 1983b; S. Froshauer and J. Beckwith, unpublished results).

A novel approach to the study of the topography of integral membrane proteins has been the use of monoclonal antibodies directed against them. Gabay *et al.* have applied this approach to the study of maltoporin (Gabay, 1982; Gabay *et al.*, 1983). Eight antibodies were used as reagents to distinguish functional and topological regions of the protein. Among these were regions localized to the cell surface, some of which were essential to transport. Others appeared to be internal to the membrane or localized to the inner surface of the outer membrane (Gabay, 1982). Two of the externally localized sites were found to reside in the carboxy terminus of the protein (Gabay *et al.*, 1983).

Functional aspects of maltoporin in *E. coli* have been extensively studied (for a review, see Nikaido *et al.*, 1980). The protein apparently binds and transports a variety of substrates including disaccharides and oligosaccharides with preferential specificity for glucosaccharides of the maltoconfiguration (Nikaido *et al.*, 1980; Luckey and Nikaido, 1980a). Illustrating the specificity of the system for disaccharides, maltose was shown to diffuse into liposomes containing the purified maltoporin 40-fold faster than sucrose. The 10 other disaccharides examined were transported at rates between those observed for these two compounds. An acetylated disaccharide, di-*N*-acetylchitobiose, was not transported.

Use of trisaccharides further illustrated the specificity of the system. Maltotriose was transported more than 100-fold faster than either raffinose or melezitose. Further studies (Luckey and Nikaido, 1980b) revealed that low concentrations of large maltooligosaccharides (of 5–7 monosaccharide units) strongly and specifically inhibited transport of smaller sugars such as glucose and maltose. The binding affinities for the penta-, hexa-, and heptamaltosaccharides were about 1 μM. Macromolecular polysaccharides such as amylose and amylopectin also were found to bind to maltoporin with high affinity (Ferenci *et al.*, 1980). From these studies it was concluded that maltoporin possesses a configurationally specific binding site for its natural maltooligosaccharide substrates.

The specificity studies described above showed that maltoporin preferentially binds and transports hexoses of the gluco configuration and oligosaccharides in which the glucosyl residues are linked α-1,4. However, these *in vitro* specificity data are insufficient to account for relative *in vivo* transport rates. For example, glucose and lactose are transported into maltoporin-containing liposomes at 300 and 10% of the rate observed for maltose, but the corresponding rates were less than 1% of the maltose influx rate when measured with intact *E. coli* cells (Luckey and Nikaido, 1980a). Since intact cells, but not liposomes, contain the maltose-binding protein (the E protein) and since extensive data are available to suggest that the periplasmic dicarboxylic acid–binding protein apparently confers specificity to porins in the outer membrane of *E. coli* (Lo and Bewick, 1981; Lo, 1979; Bewick and Lo, 1979), it seemed possible that the maltose-binding protein might bind to maltoporin and enhance its specificity for maltooligosaccharides.

Evidence for such a model was obtained by studying the transport of sugars in mutants lacking various porins but containing normal or defective maltose-binding protein (Heuzenroeder and Reeves, 1980; Ferenci and Boos, 1980). A direct interaction between the purified maltoporin and the maltose-binding protein was demonstrated by affinity chromatography (Bavoil and Nikaido, 1981). The maltose-binding protein was conjugated to Sepharose 6MB beads, and malto-porin, in a detergent extract, was passed through a column of the conjugated resin. Maltoporin was quantitatively adsorbed to the beads and could be eluted from the resin with 1 M NaCl. Sepharose beads alone, or beads to which the histidine-binding protein had been covalently linked, did not retard passage of maltoporin through the column (Bavoil and Nikaido, 1981). Moreover, other porin proteins in the *E. coli* outer membrane were not bound to the maltose-binding protein-conjugated beads. These results substantiate by direct binding the suggestion that maltoporin possesses on its inner surface a specific binding site for the maltose-binding protein. Since the maltose-binding protein binds maltodextrins with high affinity, this interaction presumably enhances the specificity of maltoporin for maltooligosaccharides in the *E. coli* cell.

To gain evidence for the physiological importance of the malto-porin–maltose-binding protein interaction to transport across the outer membrane, mutants defective for either of these two proteins were analyzed (Luckey and Nikaido, 1982, 1983; Bavoil *et al.*, 1983). Among mutant maltoporin proteins, resulting from point mutations in the *malL* gene, were two which exhibited altered transport rates and decreased affinity for immobilized maltose-binding protein (Luckey and Nikaido, 1983). Further, a defective maltose-binding protein, resulting from a point mutation in the *malE* gene and resulting in defective maltodextrin transport across the outer membrane (Wandersman *et al.*, 1979), caused a drastic decrease in the ability of the protein to bind maltoporin (Bavoil *et al.*, 1983). These results taken together provide convincing evidence that maltoporin and the periplasmic maltose-binding protein form a functionally relevant complex on the inner surface of the outer membrane.

Recently Neuhaus *et al.* (1983) have reconstituted maltoporin in planar bilayers and have examined the effect of the maltose-binding E-protein on ion conductance. In the absence of the E-protein, channel conductance was observed without discrete conductance levels. Lipopolysaccharide was not required; nor did it affect channel activity. Addition of the periplasmic maltose-binding protein induced conductance fluctuations occurring in discrete steps. The binding protein appeared to cause the channel to close. Multiple step sizes in single channel traces suggested cooperative opening and closing of up to three channels. The binding protein exhibited high affinity for maltoporin ($K_D = 1.4 \times 10^{-7} M$) and shifted the equilibrium between open and closed states in favor of the closed state. The results provide a biochemical–biophysical concept which may account for the high degree of substrate specificity in intact cells as compared with the reconstituted, isolated maltoporin.

In contrast to these results, Brass has examined the effect of a *malE* deletion mutation on the entry of lactose into the periplasm of intact *E. coli* cells via maltoporin. He observed that in a maltose-constitutive strain, the presence of the maltose-binding protein had essentially no effect on lactose entry (H. Brass, personal communication). Since maltoporin was apparently rate limiting for uptake, it was suggested that the maltose-binding protein does not control the state of the porin channel. It is possible that different experimental conditions are responsible for the contrasting results reported by Neuhaus *et al.* and by Brass.

The previous discussion dealt with the transport of maltodextrins across the outer membrane of *E. coli*. A separate and much more complex transport system is responsible for the translocation of maltodextrins across the cytoplasmic membrane. The constituents of this system and their presumed locations within the cell envelope are depicted in Figure 2.4. At the time this monograph was written, three of the four proteins of the system (E, F, and K) had been identified and their genes sequenced (Ferenci and Boos, 1980; Shuman *et al.*, 1980; Shuman and Silhavy, 1981; Shuman 1982a,b; Gilson *et al.*, 1982a,b; Bavoil *et al.*, 1980; Fowler and Zabin, 1982; M. Hofnung, personal communication; S. Froshauer, unpublished results). The hydropathy analyses of these proteins revealed that while E and K are as hydrophilic as most typical water soluble proteins, the F protein is hydrophobic (S. Froshauer, personal communication). There are two extended hydrophobic stretches near the N-terminus and four or five such stretches at the C-terminus of the F protein. All of these stretches are of sufficient length and hydrophobicity to allow passage of an α-helical segment through the membrane (S. Froshauer, unpublished results). Thus, the two ends of the protein may be embedded in the membrane, passing through it a total of six or seven times, while the middle portion extends into an aqueous environment either on the cytoplasmic or the periplasmic side of the membrane.

As shown in Fig. 2.4, the E-protein (maltose-binding protein) is localized to the periplasmic space where it can interact both with maltoporin in the outer membrane and with the F–G channel complex in the inner membrane. Both the F- and G-proteins are assumed to be integral membrane proteins which traverse the membrane and possess binding sites for maltodextrins. At least one sugar-binding site in the F–G complex has been deduced (Shuman, 1982b). This binding site is altered by a mutation in the *malG* locus (H. Shuman, personal communication). However, there is as yet no evidence for two such sites, or for the association of such binding sites with the two integral membrane constituents of the system. This suggestion (as depicted in Fig. 2.4) is based on a presumed analogy with the dicarboxylic acid transport system (Leonard *et al.*, 1981; Lo, 1977; Lo and Bewick, 1981; Bewick and Lo, 1979).

The fourth protein constituent of the system, the K-protein, is thought to be a peripheral membrane constituent of the system, bound to the cytoplasmic surface of the G-protein. This supposition is based partly on the observation that certain mutations in *malG* release the K-protein from the membrane into the cell cytosol and partly on the fact that sonication or detergent treatment releases the K-protein from the membrane in a soluble form (Shuman, 1982a,b). The hydropathy analysis of the amino acid sequence is also consistent with a peripheral membrane location. Since the G-protein is thought to bind to the K-protein, and the latter protein may function to energize the system (Gilson *et al.*, 1982a) and to regulate its activity (Saier, 1982), it might be predicted that the G-protein bears a maltodextrin binding site on the cytoplasmic side of the membrane while the F-protein possesses both an external maltodextrin binding site and the maltose-binding protein interaction site (Fig. 2.4). It should be pointed out that there is as yet no direct evidence for these postulates and that the speculative aspects of the model shown in Fig. 2.4 should serve only as a working hypothesis.

The energetics of maltose transport are still poorly understood. While the system catalyzes maltose and maltodextrin uptake and accumulates these sugars more than 100,000-fold against a concentration gradient (Boos, 1982), efflux of sugar via the system has not been demonstrated. Uptake of maltodextrins is absolutely dependent on all four proteins of the system, including the periplasmic maltose-binding protein. In the absence of any one of these four proteins, even facilitated diffusion does not occur at an appreciable rate. Dinitrophenol and other proton-conducting agents (protonophores) block both transport and accumulation of maltose. Further, after high cytoplasmic concentrations of [^{14}C]maltose are attained in *malQ* mutants (which lack amylomaltase), the addition of high concentrations of nonradioactive external maltose and/or dinitrophenol does not elicit efflux of the radioactive sugar (Boos, 1982). Slow efflux of maltose in such mutants apparently occurs after acetylation of the sugar via a different permeation route (Freundlieb and Boos, 1982).

While proton conductors block maltose uptake, suggesting a role of the proton electrochemical gradient in transport, the magnitude of this chemiosmotic driving force is insufficient to fully account for the magnitude of the maltose chemical gradient generated across the cytoplasmic membrane as the result of the action of the maltose permease. Since arsenate completely blocks maltose uptake, some investigators have suggested that a chemical form of energy may be required (Berger and Heppel, 1974; Shuman, 1982a). More recently, however, others have proposed that an oxidation–reduction process may be involved (Gilson *et al.*, 1982a; Hunt and Hong, 1983). This last suggestion is partly based on the observation that the K-protein and the respiratory NADH dehydrogenase of *E. coli* show significant sequence homology (Gilson *et al.*, 1982a,b). Interestingly, both of these proteins appear to be peripherally associated with the inner surface of the cytoplasmic membrane. At the present time, no potential energy source can be rigorously excluded as a driving source for maltose uptake, and the possibility exists that two or more energy sources are involved either in active accumulation or in the conformational regulation of the permease.

The recent work of Hunt and Hong (1983), while indecisive regarding the nature of the energy source for binding protein-dependent uptake, is of interest. These workers found that reconstituted glutamine uptake in membrane vesicles (which is driven by a binding protein–type system) occurred upon addition of pyruvate to the vesicles if NAD$^+$ was present intravesicularly. Exogenous succinate was also effective, even in the absence of NAD$^+$, although ATP and acetyl phosphate were inactive. Moreover, proton conductors and inhibitors of electron flow abolished glutamine uptake. The authors suggested that the membrane potential as well as an additional energy source may be required for the normal functioning of the periplasmic binding protein–dependent systems.

Sequence analysis of the *malK* gene suggests that the intact protein contains 370 amino acids, 42 of which are acidic (glu and asp), and 43 of which are basic (lys and arg) (Gilson *et al.*, 1982a). The distribution of polar amino acids is not random, however, and a few stretches of uninterrupted hydrophobic residues occur. The largest of these is 17 residues long, of sufficient length to pass through the membrane as an α-helix. However, the range of hydropathy values fall well within those expected for cytoplasmic or peripheral membrane proteins. The fact that this protein exhibits extensive homology to the *hisP* protein [a component of the histidine transport system which is thought to span the membrane (Gilson *et al.*, 1982b)] possibly suggests that the *hisP* protein, like the *malK* protein, is localized to the cytoplasmic surface of the membrane. It is interesting that computer analyses have revealed internal homology of an amino-terminal region of the K-protein with its own carboxyl-terminal region (Gilson *et al.*, 1982b).

While maltodextrins of greater than seven glucose residues can cross the outer *E. coli* membrane and bind to the periplasmic maltose-binding protein, maltoheptose is the largest maltodextrin which is able to cross the cytoplasmic membrane. This observation shows that neither maltoporin nor the maltose-binding protein restricts the maltodextrin specificity of the permease system. Since the latter exhibits high affinity for its substrates (K_D of about 1 μM) the need for a maltodextrin-specific porin such as maltoporin is apparent. Without it, efficient, high-affinity scavenging of these oligosaccharides would not be possible because of low transport rates across the outer membrane. The bacteria would not be able to utilize low concentrations of these sugars.

Several observations are consistent with the conclusion that the maltose permease and other binding protein-dependent systems function by a channel-type rather than by a carrier-type mechanism.

1. The maltose permease transports maltooligosaccharides of from two to seven sugar units. A mechanism in which the helical

substrate "screws" through a spiral-shaped pore is easy to envisage.

2. Binding protein–dependent systems frequently show lower temperature dependencies than do chemiosmotic symporters, which probably function by carrier-type mechanisms (Leonard *et al.*, 1981). This observation is consistent with the notion that minimal conformational changes in the overall structure of the permease would be expected for channel-mediated solute translocation.

3. Sequence homology with protein constituents of the histidine permease suggests similar mechanisms. Since both the histidine-binding protein and the lysine–arginine–ornithine (LAO)-binding protein use the same cytoplasmic membrane protein constituents for translocation of solutes into the cytoplasm and since both D- and L-amino acids are transported by the system, a solute-nonspecific pore has been proposed in this case (Higgins *et al.*, 1982; Kustu *et al.*, 1979).

4. The dicarboxylic acid permease in the *E. coli* cytoplasmic membrane possesses two distinct substrate binding sites in addition to that of the periplasmic binding protein (Leonard *et al.*, 1981; Lo, 1977). Since one of these binding sites is on the external surface of the membrane while the other is localized to the inner surface, a stationary channel, rather than a shuttling carrier, is implied. The structural similarities of this permease and the maltose permease suggest mechanistic similarities.

5. The irreversibility of maltose transport suggests tight coupling of energy consumption to solute translocation. Indeed, uncoupling of the two events would prevent maximal solute accumulation and cause energy dissipation. Furthermore, since the maltose permease cannot catalyze facilitated diffusion, either at high maltose concentrations in the absence of the maltose-binding protein or in the absence of energy, it can be assumed that the proper channel configuration for stereospecific permeation is induced both by binding of the maltose–maltose binding protein complex to the outer surface of the channel and by energization of the permease (possibly the *malK* protein) on the cytoplasmic surface of the permease complex. Such a postulate assumes that maltose induces the binding protein to change conformation (Szmelcman *et al.*, 1976; Newcomer *et al.*, 1979) and that only the complex of substrate and binding protein can open the channel. The observation that mutations in *malF* and *malG*

can allow the permease to transport maltose in the absence of the binding protein is consistent with this interpretation. The suggestion that the E-protein influences the conformation of the maltoporin channel (Neuhaus *et al.*, 1984) provides further support for this notion.

It is interesting to note that maltose-induced wild-type *E. coli* cells possess about 1000 copies of the K and F (and presumably G) proteins, while the periplasmic maltose-binding protein and maltoporin are present in about 30,000 copies per cell. Thus, lower efficiency of the less-specific transport process occurring in the outer membrane may be compensated by increased numbers of channel proteins. The number per cell and function of the proposed *malM* gene product are not known. Elucidation of the biosynthetic regulatory mechanism controlling the frequency of translation initiation should prove most interesting.

While numerous studies have demonstrated the involvement of the maltose-binding protein in maltose transport across the cytoplasmic membrane, recent reconstitution studies (Brass, 1982; Brass *et al.*, 1983) have provided virtually unequivocal proof of this role. Mutants deleted for the *malE* gene, but which synthesize the other protein constituents of the permease constitutively, cannot transport maltose. Treatment of intact cells with high concentrations of calcium in the presence of Tris buffer renders the outer membrane leaky and allows entry of the maltose-binding protein into the periplasmic space. Under these conditions, addition of the maltose-binding protein to the mutant cells restored maltose uptake, and the activity was blocked by anti-maltose-binding protein antibody or by amylopectin. Reconstituted, Ca^{2+}-treated mutant cells accumulated maltose with high affinity ($K_m = 1$–$2\ \mu M$) in a process that was absolutely dependent on metabolic energy. The reconstituted activity therefore exhibited the characteristics of the wild-type maltose permease.

While the detailed structure of the maltose-binding protein has not yet been elucidated, the application of X-ray crystallographic techniques in the laboratory of F. Quiocho has revealed the three-dimensional structures of several other solute-binding proteins (Gilliland and Quiocho, 1981; Newcomer *et al.*, 1981a,b; Vyas *et al.*, 1983; Saper and Quiocho, 1983; Fukada, *et al.*, 1983). The structures of four binding proteins from *E. coli* [those specific for D-galactose

(GBP), ʟ-arabinose (ABP), leucine, isoleucine, and valine (LIVBP) and sulfate (SBP)] are shown in Fig. 2.5. It can be seen that all such binding proteins exhibit similar structures, each having similar molecular weights and consisting of two globular domains. Each domain is made from separate polypeptide chain segments. Despite the discontinuity in the folding, the arrangements of the secondary structure in the two domains are usually similar. For example, in the arabinose-binding protein, both domains contain a six-stranded par-

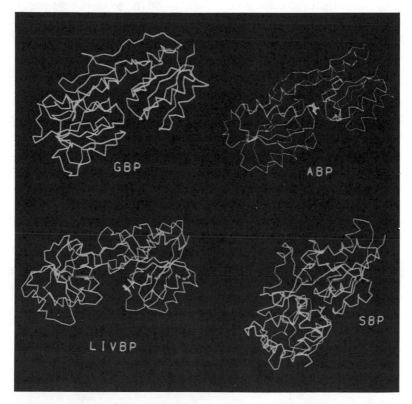

Fig. 2.5. Schematic depiction of four solute-binding proteins from *E. coli*. These proteins are the galactose-binding protein (GBP), the ʟ-arabinose binding protein (ABP), the leucine, isoleucine, valine-binding protein (LIVBP), and the sulfate-binding protein (SBP). Each of the last three proteins show the binding of their respective solutes bound in the cleft between the two domains of the protein. [Courtesy of Dr. F. A. Quiocho, Rice University, Houston, Texas.]

allel β-sheet (with the exception of the sixth antiparallel strand in one of the two domains), and these are flanked by two α-helices on either side. A C-terminal helix is shared by both domains. Although the two domains show conformational similarities, they lack significant sequence homology.

The cleft formed by the packing of the two domains of the arabinose-binding protein is predominantly lined with hydrophilic residues. The sugar-binding site is located in this cleft. When the sugar is bound, the protein undergoes a conformational change, which causes the sugar molecule to become buried in the cleft between the two lobes of the bilobate protein. All sugar hydroxyl groups are hydrogen bonded to side-chain residues in the protein as shown in Fig. 2.6. These H bonds are as follows: α- or β-OH (1) to asp-90; OH(2) to lys-10; OH(3) to asn-205, asn-232, and glu-14; and OH(4) to arg-151 and asn-232. The ring oxygen is hydrogen bonded to arg-151 (Newcomer *et al.*, 1981a; Miller *et al.*, 1983; F. Quiocho, personal communication). Lys-10, glu-14, and asp-90 are lodged in one domain while asn-205 and asn-232 are associated with the other. In Fig.

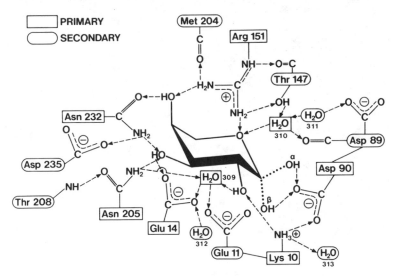

Fig. 2.6. Interaction between L-arabinose, both alpha and beta anomers, and aminoacyl residues in the binding site of the arabinose-binding protein. The schematic figure shows details of the network of hydrogen bonds between the sugar and protein. Primary residues involved in binding are boxed while secondary residues are circled. [Courtesy of Dr. F. A. Quiocho, Rice University, Houston, Texas.]

2.6, those residues which are directly hydrogen bonded to arabinose are boxed, while those which are only secondarily involved are circled.

Because of the protein conformational changes which accompany binding, the bound L-arabinose is inaccessible to the aqueous environment. These conformational changes involve a substrate-induced cleft closure in which one lobe rotates relative to the other lobe of the two-domain protein (Newcomer *et al.*, 1981b). Rotation probably occurs to the extent of 18°, about a hinge deep in the base of the sugar-binding cleft, between the two domains. A measureable consequence of this conformational change is a decrease in the radius of gyration of the protein. Because the bound arabinose is completely inaccessible to solvent, it is proposed that binding and release of the sugar must involve a rapid oscillation or "breathing" movement of the protein between the two predominant conformations.

The conformational change which is induced upon binding of sugar may be an important event permitting interaction of the binding protein with the proteins of the solute-specific transport system in the cytoplasmic membrane. Indeed, some aspects of the conformational change may trigger an opening of the transport channel which allows passage of sugar into the cytoplasm. Similar relationships may be important to the interaction of the binding protein with an outer membrane porin.

An evolutionary relationship between the different binding protein–dependent transport systems has been postulated. Substantiating this postulate is the observation that the *malK* protein component of the maltose permease shows sequence homology with the *hisP* component of the histidine permease. Moreover, each binding protein–dependent transport system which has been subjected to detailed genetic analysis consists of four proteins. Finally, extensive sequence homology has been demonstrated with some (but not other) binding proteins, and the tertiary structures of these proteins appear to show similarities. Tertiary structures may prove to be a reliable indicator of the evolutionary relatedness of two proteins which have diverged to the extent that meaningful sequence homology can no longer be demonstrated.

The discussion of maltose permeation presented in this section reveals not only the substantial progress made within the last 4 years, but also the major deficiencies in our knowledge regarding the mechansim of solute binding protein–dependent transport. At present, we have little information about the energy source(s) which drive(s) active transport via binding protein–dependent systems, and the individual functions of the four distinct proteins of the system have yet to be defined. Studies in the near future will hopefully correct these deficiencies.

3

Group Translocation Catalyzed by the Phosphoenolpyruvate: Sugar Phosphotransferase System

The phosphoenolpyruvate: sugar phosphotransferase system (PTS) is found in a wide variety of bacterial species and is involved in the transport and phosphorylation of many sugars (Dills *et al.*, 1980). The contents of this chapter will be restricted primarily to the systems found in the enteric bacteria, *Escherichia coli* and *Salmonella typhimurium*. For a discussion of the PTS in gram-positive bacteria, see Reizer *et al.*, (1985).

A.　GENERAL FEATURES OF THE PHOSPHOTRANSFERASE SYSTEM

Figure 3.1 shows three of the sequences of phosphoryl transfer reactions which are responsible for sugar group translocation in *E. coli*. In these sequences, a high energy phosphoryl group is transferred first from phosphoenolpyruvate (the initial phosphoryl donor) to the N-3 position of a histidyl residue in Enzyme I, and then from phospho-Enzyme I to the N-1 position of a histidyl residue in HPr. These two proteins are general (non-sugar-specific) cytoplasmic components of the PTS which are required for the transport of all sugar substrates of the system except for fructose (Fig. 3.1).

All additional proteins of the PTS exhibit sugar specificity. For

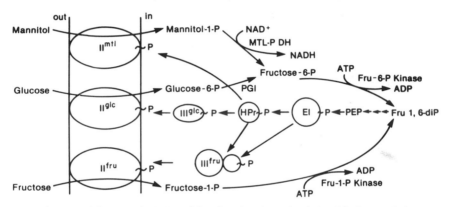

Fig. 3.1.　Schematic depiction of the phosphoryl transfer chain of the bacterial phosphotransferase system showing the enzyme constituents responsible for the transport and phosphorylation of mannitol (Mtl), glucose (Glc), and fructose (Fru). The enzymes involved in the conversion of the resultant cytoplasmic sugar phosphates to fructose 1,6-diphosphate are also shown. The abbreviations are as follows: PEP, phosphoenolpyruvate; EI, Enzyme I; HPr, heat-stable phosphocarrier protein of the PTS; III^{glc}, the glucose-specific Enzyme III of the PTS; III^{fru}, the fructose-specific Enzyme III of the PTS; II^{mtl}, II^{glc}, and II^{fru}, the mannitol-specific, the glucose-specific, and the fructose-specific Enzymes II of the PTS, respectively. The Enzymes II are integral membrane proteins which function as the sugar permeases. The remaining enzymes of the PTS depicted are soluble or peripherally associated with the membrane. Each protein must be phosphorylated in sequence in order for group translocation of the sugar to occur. The phosphotransferase system is depicted as being integrated into the glycolytic pathway for the generation of phosphoenolpyruvate. Additional enzymes depicted are: mannitol-1-P dehydrogenase (MTL-P DH), fructose-6-P kinase, phosphoglucoisomerase (PGI), and fructose-1-P kinase.

example, glucose transport involves two glucose-specific proteins, a soluble (or peripheral membrane) Enzyme III^{glc} of 20,000 molecular weight, phosphorylated at the expense of phospho-HPr on the N-3 position of a histidyl residue in the protein (see Chapter 4, Section C) as well as an integral membrane Enzyme II^{glc} of 40,000–50,000 molecular weight (Erni *et al.*, 1982). Enzyme II^{glc} is phosphorylated at the expense of phospho-Enzyme III^{glc} (Begley *et al.*, 1982; B. Erni, unpublished results), possibly on the N-1 position of a histidyl residue (E. B. Waygood, unpublished results). Finally, phospho-Enzyme II^{glc} donates its phosphoryl moiety to glucose. The sugar is concomitantly translocated and phosphorylated so that glucose 6-phosphate is the product released from the permease into the cytoplasm (Fig. 3.1).

Mannitol phosphorylation involves only a single sugar-specific enzyme, the integral membrane Enzyme II^{mtl}. The product of mannitol phosphorylation, cytoplasmic mannitol 1-phosphate, is oxidized to fructose 6-phosphate in a reaction catalyzed by mannitol-1-phosphate dehydrogenase (Fig. 3.1). The initiation of fructose metabolism, the third sugar whose transport-coupled phosphorylation is depicted in Fig. 3.1, is more complex than that of either glucose or mannitol (Waygood *et al.*, 1979; E. B. Waygood, 1980; and personal communication). Transfer of phosphate from phospho-Enzyme I can probably proceed either to HPr or to a fructose-inducible, HPr-like protein, designated FPr (MW = 8000), which in the cell may be complexed with the fructose-specific Enzyme II–Enzyme III pair. Phospho-HPr or phospho-FPr then probably catalyzes Enzyme III^{fru} phosphorylation, and phospho-Enzyme III^{fru} may phosphorylate Enzyme II^{fru} in the membrane. Whether phospho-HPr can donate its phosphoryl group to FPr has not been determined. The presumed (but not proven) sequence of phosphoryl transfer is therefore PEP → Enzyme I → HPr or FPr → Enzyme III^{fru} → Enzyme II^{fru} → fructose.

In addition to glucose, mannitol, and fructose, the PTS phosphorylates several other sugars in *E. coli*. These sugars include hexoses of the gluco configuration such as *N*-acetylglucosamine, phosphorylated by the Enzyme II^{nag}, β-glucosides, phosphorylated by the Enzyme II^{bgl}, and mannose, phosphorylated by the Enzyme II^{man}–Enzyme III^{man} pair. This last enzyme complex shows low specificity for its sugar substrates, phosphorylating glucose, mannose, 2-deoxyglucose, *N*-acetylglucosamine, glucosamine, *N*-acetylmannosamine, mannosamine, fructose, and methyl α-glucoside, with decreasing affinity in this order (Rephaeli and Saier, 1980a). The Enzyme III^{man}, which is partially soluble and partially membrane bound, appears to have a molecular weight of about 33,000 (E. B. Waygood, personal communication), while the Enzyme II^{man} has been reported to have a molecular weight of about 36,000 (Kundig and Roseman, 1971) or 25,000 (B. Erni, personal communication). Also phosphorylated by distinct phosphotransferases are the hexitols, mannitol [phosphorylated by the Enzyme II^{mtl} (Lee et al., 1981)], glucitol [phosphorylated by the Enzyme II^{gut}–Enzyme III^{gut} pair (Grenier et al., 1985; Sarno et al., 1984)] and galactitol [phosphorylated by the Enzyme II^{gat} (Lengeler, 1975b, 1977)]. Enzymes III specific for β-glucosides and galactitol have not yet been identified and may or may not exist (Dills et al., 1980). Extensive evidence has led to the conclusion that a mannitol Enzyme III is not present in E. coli or S. typhimurium (Jacobson et al., 1979; 1983b; Lee et al., 1981). Further, unpublished results from the laboratory of E. B. Waygood suggest that no Enzyme III participates in the phosphorylation of N-acetylglucosamine. Instead, a single, sugar-specific, integral membrane Enzyme II^{nag} (MW = 65,000) catalyzes the uptake and phosphorylation of this sugar. Thus, eight Enzymes II (glucose, mannose, N-acetylglucosamine, β-glucoside, fructose, mannitol, glucitol, and galactitol) and four Enzymes III (glucose, mannose, fructose, and glucitol) have been identified in E. coli and S. typhimurium to date. Table 3.1 summarizes the properties of these proteins.

The list of PTS proteins included in Table 3.1 is undoubtedly incomplete. Jin and Lin (1984) have recently shown that dihydroxyacetone is transported and phosphorylated by a phosphotransferase-mediated mechanism. Evidence was presented which suggested that synthesis of the triose-specific component(s) of the system was induced by growth in the presence of dihydroxyacetone and that the system was distinct from the previously characterized Enzyme II–III pairs. Additionally Marechal (1984) has published convincing evidence showing that trehalose enters E. coli and S. typhimurium cells via the phosphotransferase system. In this work, the product of trehalose transport, trehalose 6-phosphate, was identified, and a trehalose-inducible trehalose-6-phosphate hydrolase was detected which generated glucose and glucose 6-phosphate as hydrolytic products. Because phosphoenolpyruvate but not ATP could serve as the phosphoryl donor for trehalose phosphorylation, a PTS-mediated mechanism appears established. The nature of the

TABLE 3.1

Properties of Phosphoryl Transfer Proteins of the PTS in *E. coli*[a]

Protein[b]	Molecular weight[c]	Purified to apparent homogeneity	Phosphorylation demonstrated	Position of phosphorylation
Enzyme I	60,000	+	+	N-3 his
HPr	9,000	+	+	N-1 his
FPr	8,000		+	N-1 his
Enzyme III[gut]	16,000	+	+	N-3 his
Enzyme III[glc]	20,000	+	+	N-3 his
Enzyme III[fru]	40,000		+	N-3 his
Enzyme III[man]	34,000		+	N-3 his
Enzyme II[mtl]	68,000	+	+	N-3 his and N-1 his
Enzyme II[gut]	45,000		(+)	
Enzyme II[gat]				
Enzyme II[glc]	45,000	+	+	N-1 his
Enzyme II[fru]				
Enzyme II[man]	25,000		+	
Enzyme II[nag]	65,000		+	N-3 his and N-1 his
Enzyme II[bgl]				

[a] This list summarizes the known proteins of the PTS in *E. coli*. Extensive evidence argues against the existence of an Enzyme III[mtl] in *E. coli*. Enzymes III specific for galactitol, N-acetylglucosamine, and β-glucosides have not yet been identified, but the possibility of their existence has not been rigorously excluded.

[b] The abbreviations used are glc, glucose; mtl, mannitol; gut, glucitol; gat, galactitol; fru, fructose; man, mannose; nag, N-acetylglucosamine; bgl, β-glucosides.

[c] Most of the molecular weights were determined by SDS gel electrophoresis with the following exceptions: Enzyme I, sedimentation equilibrium and gel filtration under denaturing conditions; HPr, amino acid sequence; Enzyme II[mtl], amino acid sequence deduced from the gene sequence. The molecular weight of Enzyme I estimated by SDS gel electrophoresis is 65,000 while that of the Enzyme II[mtl] is 60,000. Although the *E. coli* Enzyme III[gut] has an apparent molecular weight of 16,000, that of *S. typhimurium* has an apparent molecular weight of 13,000 by SDS-polyacrylamide gel electrophoresis. This apparent difference in molecular weight between these two bacterial species represents the only significant molecular weight difference so far observed for any one enzyme of the PTS (F. C. Grenier, unpublished results; E. B. Waygood, unpublished results). The molecular weight for Enzyme II[man] has been reported to be 36,000 by Kundig and Roseman (1971) and 25,000 by B. Erni, (personal communication).

sugar-specific proteins which mediate the transport and phosphorylation of dihydroxyacetone and trehalose has yet to be determined.

B. BIOCHEMICAL AND GENETIC NOMENCLATURE OF
THE PTS

The biochemical and genetic nomenclature I recommend for the
proteins of the PTS corresponds to that used in previous publications from our laboratory (Saier and Stiles, 1975; Saier, 1977; Dills
et al., 1980; Grenier *et al.,* 1984) and several other laboratories.
An Enzyme Commission number has been assigned to Enzyme I
(EC 2.7.3.9), and EC 2.7.1.69 has been assigned collectively to the
Enzymes II (or Enzyme II–Enzyme III pairs) of the PTS. The
known PTS enzymes and their properties in *E. coli,* their estimated
molecular weights, and their positions of phosphorylation if known
are all included in Table 3.1.

The preferred genetic nomenclature for the genes encoding the
PTS proteins (Table 3.2) is as recommended previously (Demerec *et
al.,* 1966; Lin, 1970; Lengeler, 1975a; Saier *et al.,* 1976b; Rephaeli
and Saier, 1980a). Thus, the genes which encode HPr and Enzyme I
are designated *ptsH* and *ptsI,* respectively (Table 3.2). These genes
comprise an operon at 52 min on the *E. coli* linkage map (Bachmann,
1983). The gene which encodes Enzyme III^{glc} is designated *crrA* for
*c*arbohydrate *r*epression *r*esistance, by virtue of the role of this protein as a central regulatory protein controlling inducer uptake and
adenylate cyclase activity (see Sections C in Chapters 4 and 5, respectively). This gene maps adjacent to the *pts* operon in *S. typhimurium* and *E. coli* (Saier and Roseman, 1976a; Burd *et al.,* 1980;
Meadow *et al.,* 1982a,b).

Genes encoding the Enzymes II and Enzymes III of the PTS are
designated by three small letters which indicate the sugar whose
utilization is affected and the capital letter "A" or "B" which indicates the fact that the gene encodes an Enzyme II or III, respectively, Each different mutant allele within a particular regulon
should be assigned a different allele number. Thus, *mtlA61* is a
correct designation for a mutant allele in the structural gene for the
Enzyme II^{mtl}; *gutA153* indicates a mutant allele which gives rise to a
defective Enzyme II^{gut}, and *gutB152* indicates a mutant allele with a
defect in the Enzyme III^{gut} structural gene (Sarno *et al.,* 1984). Al-

TABLE 3.2

Structural Genes of Operons Encoding the Proteins of the PTS in *E. coli*

Structural gene (recommended nomenclature)[a,b]	Alternative genetic designation[d]	Protein encoded by the gene (see Table 3.1)	Gene position
ptsI	—	Enzyme I	52
ptsH	—	HPr	52
crrA	*crr*	Enzyme III^{glc}	52
glcA	*ptsG*	Enzyme II^{glc}	24
manA	*ptsM*	Enzyme II^{man}	40
manB[c]	*ptsM*	Enzyme III^{man}	(40)
pmi	*manA*	Phosphomannose isomerase	36
nagA	*nagE*	Enzyme II^{nag}	16
nagC	*nagA*	N-Acetylglucosamine 6-P deacetylase	16
nagD	*nagB*	Glucosamine-6-P deaminase	16
bglA	*bglC*	Enzyme II^{bgl}	83
bglC	*bglA*	P-β-glucosidase A	—
bglD	*bglB*	P-β-glucosidase B	—
mtlA	—	Enzyme II^{mtl}	81
mtlD	—	Mannitol-1-P dehydrogenase	81
gutA	*srlA*	Enzyme II^{gut}	58
gutB	—	Enzyme III^{gut}	58
gutD	*srlD*	Glucitol-6-P dehydrogenase	58
gatA	—	Enzyme II^{gat}	47
gatD	—	Galactitol-6-P dehydrogenase	47
fruA	*ptsF*	Enzyme II^{fru}	46
fruB[c]	—	Enzyme III^{fru}	(?)
fruH[c]	—	FPr	(?)
fruK	*fpk*	Fructose-1-P kinase	46

[a] The recommended designations are in accordance with the recommendation of Demerec et al. (1966), Lin (1970), Lengeler (1975), Saier et al. (1976b), Rephaeli and Saier (1980a,b), and Sarno et al. (1984).

[b] Proposed genetic designations for genes encoding possible Enzymes III of the PTS which have not (as yet) been identified are *gatB* (Enzyme III^{gat}), and *bglB* (Enzyme III^{bgl}).

[c] These genes have not yet been characterized.

[d] These designations were used by Bachmann (1983). Map positions and appropriate references are to be found in Bachmann (1983).

ternative genetic and biochemical designations for the phosphotransferase system used by various laboratories in the past have been summarized previously (Rephaeli and Saier, 1980a).

Table 3.2 summarizes the known genes encoding the proteins of the PTS in *E. coli* as well as the designations for a few hypothetical genes (gene positions in parentheses) which must code for the few known proteins of the system for which mutants are not yet available. Genes encoding carbohydrate-specific catabolic enzymes present within the same operons are also indicated. Since all of these proteins except the Enzyme III^{glc} are inducible in at least some strains of *E. coli* and *S. typhimurium* (Kornberg and Reeves, 1972; Saier *et al.*, 1976a, 1977; Rephaeli and Saier, 1980b), it is clear that regulatory genes controlling expression of these operons must exist. Some of these have been identified (Bachmann, 1983). The recommended genetic designations, those currently in use (see Bachmann, 1983) and the map positions of the genes are included in the table.

C. PROPERTIES AND MECHANISM OF ACTION OF ENZYME I

The preparation of homogeneous Enzyme I from *E. coli* or *S. typhimurium* has been reported from three separate laboratories (Robillard *et al.*, 1979; Waygood and Steeves, 1980; Weigel *et al.*, 1982a). Techniques such as hydrophobic chromatography employing octylsepharose (Robillard *et al*, 1979), ion-exchange chromatography, and gel filtration (Waygood and Steeves, 1980; Weigel *et al.*, 1982a) were used. Purification by ion exchange on DEAE cellulose was facilitated by preincubation of Enzyme I with phosphoenolpyruvate and Mg^{2+} to generate the phosphorylated form of the enzyme, thus enhancing the negative charge of the protein. Purification by gel filtration took advantage of the temperature-dependent association of Enzyme I monomers. While the enzyme eluted from the column as a monomer (apparent MW = 67,000) at 4°C in the presence of EDTA and the absence of Mg^{2+}, it eluted as a dimer (apparent MW = 135,000) at 23°C in the presence of 10 mM MgCl$_2$. The dimeric species appeared to be a homodimer of identical subunits since the amino-terminal sequence revealed a single amino acid at each of the first 17 positions. The amino acid compositions of the proteins from both *E. coli* and *S. typhimurium* were reported (Waygood and Steeves, 1980; Weigel *et al.*, 1982a), and the molecular weight, estimated by a sedimentation equilibrium experiment or by gel filtration under denaturing conditions, was near 60,000 (Kukuruzinska *et al.*, 1982).

The associative properties of the enzyme have been examined by several investigators. Reaction rate versus enzyme concentration showed sigmoidal behavior both for the phosphorylation of sugar in the presence of the other proteins of the PTS and for phosphoenolpyruvate:pyruvate exchange, measured in the absence of the other PTS proteins (Saier et al., 1980). Moreover, lag periods for methyl α-glucoside phosphorylation were noted, and the duration of the lag phase decreased as the Enzyme I concentration increased (Misset et al., 1980). If Enzyme I was preincubated with Mg^{2+} and phosphoenolpyruvate at 37°C, the lag period was eliminated. Thus, in agreement with the observation of Weigel et al., higher temperature in the presence of Mg^{2+} and phosphoenolpyruvate appears to promote dimerization, and only the dimeric form, complexed with Mg^{2+}, appears to be phosphorylated at the expense of phosphoenolpyruvate. The metal ion was reported to enhance the stability of the dimer approximately 10-fold, and the metal bound more tightly to the phosphorylated form of the enzyme than the free form (Hoving et al., 1982). On the basis of the salt dependencies of the association process, it was suggested that hydrophobic interactions play a role in subunit association. The possibility of higher oligomeric species (trimers, tetramers, etc.) has not been completely ruled out (Kukuruzinska et al., 1982). However sedimentation equilibrium and sedimentation velocity studies have so far failed to reveal such species (E. B. Waygood, unpublished results).

The kinetic properties of Enzyme I-catalyzed reactions have been reported. The pH optimum is between 7.0 and 7.5 both for sugar phosphorylation and for the phosphoenolpyruvate:pyruvate exhange reaction (Saier et al., 1980; Waygood and Steeves, 1980). Both phosphoenolpyruvate:pyruvate exchange (Saier et al., 1980) and the phosphoenolpyruvate:HPr phosphotransfer reaction (Weigel et al., 1982b; Waygood and Steeves, 1980) displayed a Ping-Pong Bi-Bi mechanism. The different assay procedures yielded K_m values for phosphoenolpyruvate between 0.2 and 0.5 mM, K_m values for HPr between 5 and 10 μM, and a K_m value for pyruvate of 2 mM. K_a values for Mg^{2+} were reported to be between 0.5 and 2 mM while those for Mn^{2+} and Co^{2+} were at least an order of magnitude lower. Finally, employing the exchange assay, phosphoenolpyruvate was shown to form an abortive complex with phospho-Enzyme I (K_i = 2.5 mM) while pyruvate formed an abortive complex with free Enzyme I ($K_i \cong 20$ mM) (Saier et al., 1980). These results provide a reasonably complete kinetic characterization of the interactions of Enzyme I with its substrates, products, and activating metal ions.

It should be noted that Misset and Robillard (1982) did not observe Ping-Pong kinetics as did the other investigators (see also Hoving *et al.*, 1981). An interpretation of their results is presented in a recent review (Robillard, 1982). Another conflict involves the number of metal ions and phosphoryl groups present in dimeric phospho-Enzyme I. Although Hoving *et al.* (1982) reported that one metal ion (Mg^{2+} or Mn^{2+}) binds to dimeric Enzyme I and only one phosphoryl group is incorporated per dimer, Weigel *et al.* (1982b) reported that two phosphoryl groups can be maximally incorporated into the Enzyme I dimer. The latter investigators showed that transfer of the phosphoryl groups from phospho-Enzyme I to HPr occurs in the absence of a divalent cation and in the presence of 20 mM EDTA. This result suggests that the monomeric form of phospho-Enzyme I can interact with monomeric HPr to transfer its phosphoryl moiety in the absence of a divalent cation.

The phosphoryl donor–acceptor specificity of Enzyme I has been examined by studying the phosphoenolpyruvate : α-keto acid phosphotransfer exchange reaction catalyzed by this enzyme (Saier *et al.*, 1980). α-Ketobutyrate replaced pyruvate at about 20% the efficiency of the natural substrate. β-Hydroxypyruvate was another phosphoryl acceptor (4% efficiency relative to pyruvate), but α-ketovalerate and α-ketocaproate were virtually inactive. A recent report has shown that proton transfer in the phosphorylation of Enzyme I occurs stereospecifically. While Z-phosphoenolbutyrate is a substrate, the E-isomer is not (Hoving *et al.*, 1983). Proton transfer to the C-3 position of Z-phosphoenolbutyrate must therefore occur stereospecifically. It is interesting to note that the pyruvate analogue, α-ketobutyrate, has recently been shown to be a physiologically relevant regulator of PTS function under certain conditions of growth (Daniel *et al.*, 1983). When this keto acid was generated *in vivo* from threonine by the action of threonine deaminase, the appearance of cytoplasmic α-ketobutyrate was accompanied by aspartate starvation, an elevation in the endogenous ppGpp pools, and cessation of PTS-sugar utilization. An understanding of the keto acid specificity of Enzyme I may therefore provide an explanation for adaptive behavior of the bacteria in response to environmental changes.

D. PROPERTIES OF HPr

The small phosphocarrier protein, HPr (heat-stable, histidine-containing protein), has been purified to homogeneity and sequenced from four bacterial species: *Salmonella typhimurium, Staphylococcus aureus*, Streptococcus faecalis, and *Bacillus subtilis*. The proteins from *E. coli* and *S. typhimurium* are very similar in amino acid composition and NMR spectra, and they exhibit full enzymatic cross reactivity (Anderson *et al.*, 1971; Weigel *et al.*, 1982a; Lee *et al.*, 1982; W. Hengstenberg, unpublished results). These proteins may therefore be identical or nearly identical. By contrast, the protein from *S. typhimurium* does not substitute well for those from the three gram-positive bacterial species although the latter three are enzymatically interchangeable. Antibodies directed against the *S. aureus* HPr inhibited the function of the HPr proteins from streptococcal species and from *B. subtilis* (W. Hengstenberg and J. Reizer, unpublished observation). It is interesting to note that the *Salmonella* protein exhibits only 30% sequence homology with that from *S. aureus*. For a detailed discussion of the proteins of the phosphotransferase system from gram-positive bacteria, see Reizer *et al.* (1986).

Roossien *et al.* (1979) reported that purified HPr from *E. coli* was heterogeneous. Based on ^{1}H-NMR analyses, it was concluded that a tyrosyl residue was located in a single position in the interior of 50% of the HPr molecules from *E. coli*. Anderson *et al* (1971) had previously noted the absence of tyrosine in *E. coli* HPr, and the same observation has been reported for the *S. typhimurium* enzyme (Weigel *et al.*, 1982a). This conflict could have arisen from the presence of a second HPr species or to impurities in the preparation of Roossien *et al.* The latter explanation appears to be correct. When an HPr preparation obtained from Robillard was subjected to iodination by the Chloramine-T method (Greenwood *et al.*, 1963) and subsequently subjected to gel electrophoresis in SDS, the radioactive iodinated proteins, labeled on tyrosyl residues, were of higher apparent molecular weights, migrating at positions different from that of HPr (F. C. Grenier, R. Yang, and M. H. Saier, Jr., unpublished results). These results suggest that the tyrosyl residues in the HPr preparation of Roossien *et al.* were present in proteins other than HPr.

A second report from the Robillard laboratory (Dooijewaard *et al.*, 1979a) claimed that HPr and an α-1,6-glucan co-purified and that these two macromolecules therefore possess high affinity for each other. Since other investigators have obtained carbohydrate-free preparations of HPr (Beneski *et al.*, 1982; Weigel *et al.*, 1982a; J. Deutscher, personal communication) and since the HPr preparation obtained by Roossien *et al.* was impure (see the preceding paragraph), the significance of the presence of the glucose homopolymer in their preparation remains to be ascertained.

The amino acid sequence of HPr from *S. typhimurium* was determined (Weigel *et al.*, 1982c). It contained 84 amino acyl residues, 2 of which were histidyl residues. One histidyl residue at position 15 could be phosphorylated, while the other at position 75 could not. Phosphorylation occurred at the N-1 position of histidyl residue 15, and NMR studies suggested that the protein and its phosphorylated derivative could exist in two distinct conformations (Dooijewaard *et al.*, 1979b). The results leading to this last suggestion are, however, open to interpretation (W. Hengstenberg, personal communication).

Based on published methods (Chou and Fasman, 1978; Schiffer and Edmundson, 1967), the secondary structure of *Salmonella* HPr was predicted (Weigel *et al.*, 1982c). The predicted structure had 55% α-helix, 32% β-structure, and 23% random coil and resembled that previously predicted for the *S. aureus* protein (Beyreuther *et al.*, 1977). Interpretations of the circular dichroism spectrum of the protein were in accord with this prediction. A model for the protein, with the active histidyl residue in a region of the polypeptide assuming a random coil on the surface of the globular structure, was presented (Weigel *et al.*, 1982c). Preliminary X-ray diffraction analyses of *E. coli* HPr crystals led to a calculation of the unit cell dimensions (Delbaere *et al.*, 1982) and showed that the active histidyl residue is out on a loop, away from the bulk of the protein. However, the predictions of secondary and tertiary structure, deduced by computer analysis of the primary structure (Weigel *et al.*, 1982c) were not, in general, verified (L. Delbaere and E. B. Waygood, personal communication).

Recently, radioactive HPr as well as HPr derivatives containing fluorescent and electron paramagnetic resonance probes have been synthesized (Grill *et al.*, 1982; Hildenbrand *et al.*, 1982). The modified proteins, derivatized at the N-terminus, were fully active in the sugar phosphorylation reaction. These derivatives should be valuable for studying the catalytic properties and protein–protein interactions of the PTS components.

The *ptsH* gene from *E. coli* has been cloned into the high copy number plasmid, pBR322 (Lee *et al.*, 1982). HPr was synthesized in large amounts by a *recA* strain of *E. coli* harboring this plasmid. This important advance should facilitate preparation of large amounts of pure HPr and allow physicochemical studies which otherwise would have been impractical. The recent cloning of the entire *pts* operon as well as the *crrA* gene, encoding the glucose Enzyme III is another major technical advance which should facilitate genetic and biochemical analysis of these energy coupling proteins of the PTS (Bitoun *et al.*, 1983).

For many years it has been recognized that fructose induces the synthesis of a protein which can substitute for HPr (Saier *et al.*, 1970, 1976b). Thus, all *ptsH* (HPr⁻) mutants of *E. coli* and *S. typhimurium* are capable of fructose utilization although *ptsI* (Enzyme I⁻) mutants are not. These same *ptsH* mutants can not utilize hexitols or *N*-acetylglucosamine at appreciable rates although mannose and glucose are utilized slowly (Saier *et al.*, 1970). Growth of a *ptsH* mutant in the presence of fructose induces the capacity of the cells to utilize and phosphorylate all sugar substrates of the PTS at rates which are low relative to the wild-type strain. Growth of fructose-pregrown cells on sugars such as mannitol and *N*-acetylglucosamine occurs for only a few generations presumably because these sugars cannot induce synthesis of the HPr-like protein.

The HPr mutants were also found to utilize fructose at rates which were about one-half of those of the parental strains. These results suggested that the fructose (*fru*) regulon might encode an HPr-like protein which can partially substitute for HPr. Passage of extracts derived from fructose grown *ptsH* mutants of *S. typhimurium* through gel filtration columns suggested that a protein complex (MW of about 50,000) was responsible for the activity, but that dissociation of the complex resulted in substantial loss of activity (M. H. Saier, Jr. and S. Roseman, unpublished results).

The identities of the protein constituents of this complex have recently been determined (E. B. Waygood, 1980, and personal communication). It apparently consists of two proteins, one of molecular weight equal to 8000 and termed FPr (fructose-inducible HPr), the other of molecular weight equal to 40,000 and termed Enzyme III^fru (Table 3.1). Both proteins in the complex are phosphorylated in the presence of phosphoenolpyruvate and Enzyme I. While a histidyl residue in FPr is phosphorylated on the N-1 position (as is the case for HPr), the N-3 position of a histidyl residue in the Enzyme III^{fru} is phosphorylated (as is true for the other known Enzymes III of the PTS). It is interesting to note that the HPr proteins from gram-positive bacteria, like FPr, are smaller than *E. coli* HPr.

E. PROPERTIES AND MECHANISM OF ACTION OF ENZYME II^{mtl}

The Enzymes II of the PTS are by far the most interesting enzyme constituents of the system because these proteins catalyze vectorial group translocation of their sugar substrates across the membrane. In fact, they catalyze two group translocation processes: unidirectional transport, coupled to phosphoenolpyruvate-dependent sugar phosphorylation, and bidirectional transport, coupled to sugar phosphate-dependent sugar phosphorylation. A report that the Enzymes II of the PTS can catalyze nonvectorial phosphorylation in intact cells (Saier and Schmidt, 1981) proved to be erroneous (Robillard and Legeveen, 1982; Saier and Leonard, 1983). While this reaction can be demonstrated in membrane vesicle preparations, with membrane fragments and with solublized Enzyme II^{mtl} suspended in detergent micelles, nonvectorial phosphorylation does not appear to occur appreciably in intact cells.

Several reports suggest that the Enzymes II can facilitate transport of free sugar across the membrane, unidirectionally from the inside of the cell to the external medium in the absence of phosphorylation [see Reizer and Saier (1983) for a discussion of the evidence concerning this possibility]. Possibly the binding of free sugar to the sugar-P binding site on the cytoplasmic side of the Enzyme II opens the channel in the enzyme through which the sugar can pass. This hypothesis has not yet been rigorously tested. The Enzymes II may catalyze the slow uptake of free sugar, but at rates which are substantially slower than the efflux rates (Lengeler and Steinberger 1978a,b; Reizer and Saier, 1983). There is also substantial evidence suggesting that these proteins function as sugar-specific chemoreceptors (Lengeler *et al.*, 1981; Pecher *et al.*, 1983) and that at least some of them may function in the transcriptional regulation of certain operons (see Chapter 7) (Saier, 1980; Saier and Leonard, 1983). Partial genetic dissection of the different functions of the Enzyme II^{mtl} has been accomplished (Leonard and Saier, 1981; G. Tenn, D. Printz, and M H. Saier, Jr., unpublished results).

The kinetics of Enzyme II^{lac}-catalyzed transfer of phosphate from phospho-Enzyme III^{lac} to lactose, employing partially purified enzymes from *Staphylococcus aureus*, was shown to occur by a Bi-Bi sequential mechanism (Simoni *et al.*, 1973). By contrast, the transfer of phosphate from phospho-HPr to mannitol, catalyzed by purified mannitol Enzyme II of *E. coli* exhibited Ping-Pong kinetics (C. A. Lee and M. H. Saier, Jr., unpublished results). While the former result suggested that phospho-Enzyme III^{lac} and lactose must bind simultaneously to the enzyme before reaction can occur, the latter results argue in favor of a phosphorylated Enzyme II intermediate. Phosphorylated Enzyme II^{mtl}, derivatized on the N-3 position of a histidyl residue, has been tentatively identified (E. B. Waygood, personal communication). Since this membrane-bound enzyme functions without the participation of an Enzyme III (Lee *et al.*, 1981), its phosphorylation may be analogous to the phosphorylation of an Enzyme III in the group translocation of other sugars. If the Enzymes II are in fact phosphorylated, as appears to be the case (E. B. Waygood, personal communication; Begley *et al.*, 1982; B. Erni, unpublished results), the Enzyme II^{mtl}, which may be functionally equivalent to an Enzyme II–Enzyme III pair, may be phosphorylated twice as part of the normal phosphoryl transfer sequence. Further experiments will be required to test this speculation.

Evidence for the phosphorylation of Enzyme II^{glc} in *E. coli* has been obtained with the use of chiral phosphoenolpyruvate (Begley *et al.*, 1982; Knowles, 1980). This conclusion was deduced from the fact that transfer of phosphate from phosphoenolpyruvate to glucose 6-phosphate occurred with inversion of configuration. This result implies an odd number of phosphoryl transfer reactions, assuming that each transfer inverts the configuration of groups about the phosphorous atom. Since Enzyme I, HPr, and Enzyme III^{glc} are all known to be phosphorylated as part of the phosphotransfer sequence, Enzyme II is presumably phosphorylated before transfer of phosphate can occur to glucose. This conclusion assumes that each of the soluble energy-coupling enzymes is phosphorylated on a single histidyl residue. Recently the direct phosphorylation of Enzyme II^{glc} has been demonstrated (B. Erni, unpublished results). Phosphorylation with [^{32}P]phosphoenolpyruvate depended on Enzyme I, HPr, and Enzyme III^{glc}, as expected.

Recent preliminary evidence has argued in favor of the conclusion that a glucitol Enzyme II-Enzyme III pair is required for glucitol phosphorylation (Grenier *et al.*, 1985). The former protein appears to have a molecular weight of 45,000; the latter protein has a molecular weight near 15,000. Enzyme III^{gut} phosphorylation by phospho-HPr on the N-3 position of a histidyl residue in the Enzyme III has been demonstrated. The limited amount of evidence presently available suggests that the Enzymes II of the PTS are first phosphorylated at the expense of phospho-HPr or a phospho-Enzyme III and that the phosphorylated Enzymes II catalyze group translocation of the sugar across the membrane concomitant with transfer of the phosphoryl moiety to sugar.

Kinetic analyses of the nonvectorial chemical reaction, corresponding to the bidirectional transphosphorylation reaction ([^{14}C]sugar$_{out}$+sugar-P$_{in}$→[^{14}C]sugar-P$_{in}$+sugar$_{out}$), were conducted with butanol–urea-extracted membranes from *S. typhimurium* (Rephaeli and Saier, 1978, 1980a). In these studies the glucose and mannose phosphotransferases were examined. The Lineweaver-Burk plots of the kinetic data gave intersecting lines, characteristic of a Bi-Bi Sequential mechanism. Further studies with dead-end inhibitors led to the suggestion that the mechanism of the transphosphorylation reaction is rapid equilibrium, random, Bi-Bi Sequential.

Three investigators have independently reported that Enzyme II–catalyzed transphosphorylation reactions occur by a mechanism yielding Ping-Pong kinetics. Thus, Perret and Gay (1979) examined the fructose PTS in *Bacillus subtilis*, Hüdig and Hengstenberg (1980) studied the glucose PTS in *Streptococcus faecalis*, and Robillard (1982) reported a kinetic study of the glucose and mannose transphosphorylation processes employing Enzyme II^{glc} and Enzyme II^{man} from *E. coli*, respectively. Recently, the work with *Streptococcus* has been discredited for technical reasons (W. Hengstenberg, personal communication). Further, Robillard has found that alterations in the experimental conditions employing the *E. coli* preparations gave rise to sequential kinetics similar to those reported by Rephaeli and Saier (1978, 1980a; G. T. Robillard, personal communication). It is possible that the conditions of the reaction influence the conclusions of these kinetic studies.

Sequential kinetics usually are interpreted in terms of a mechanism in which both substrates (sugar and sugar phosphate in this case) bind simultaneously to the enzyme surface. Such a mechanism would be possible if the Enzyme II monomer possesses two topologically distinct binding sites for sugar and sugar phosphate, respectively, or if two subunits of the Enzyme II interact to form a functional dimer which is responsible for bidirectional group translocation. Rephaeli and Saier (1980a) argue against two topologically distinct binding sites for sugar and sugar phosphate. They point out that the relative affinities of the Enzyme II^{man} for its various sugar substrates were the same (a) when transport was studied in intact cells, (b) when phosphoenolpyruvate-dependent phosphorylation of the sugars was studied *in vitro* in the presence of the energy-coupling proteins of the system, and (c) when the different sugars were examined in the transphosphorylation reaction. More importantly, the relative affinities of the homologous series of sugar phosphates were the same both when the transphosphorylation reaction was studied *in vitro* and when sugar phosphate-stimulated uptake of [^{14}C]sugar was studied in *E. coli* membrane vesicles. These results were interpreted in terms of a single alterable binding site which shuttles between two states, one possessing a sugar recognition site on the external surface of the membrane, the other possessing a sugar phosphate recognition site exposed to the cytoplasmic surface of the membrane. In support of this notion was the observation that transphosphorylation, but not the phosphoenolpyruvate-dependent phosphorylation of sugar, is subject to strong substrate inhibition at high concentrations of sugar and sugar-P (Saier *et al.*, 1976a,b). If a single binding site for substrate does, in fact, shuttle between two states, exposed alternately to the two sides of the membrane, then a mechanism involving the simultaneous binding of both substrates to the enzyme surface would require subunit interactions.

In order to determine the mechanisms of uni- and bidirectional group translocation catalyzed by the mannitol Enzyme II, the enzyme was extracted from the membrane with deoxycholate in the presence of high salt and purified by hydrophobic chromatography and hydrophobic ion exchange (Jacobson *et al.*, 1979, 1983b). The purified protein gave a single band by SDS-polyacrylamide gel electrophoresis. It had a subunit molecular weight of 60,000 by this criterion, although the estimated molecular weight, based on the amino acid sequence deduced from the gene sequence, was 67,893 (Lee and Saier, 1983a,b). The purified protein catalyzed both the mannitol 1-phosphate–dependent transphosphorylation reaction and the phosphoenolpyruvate-dependent phosphorylation of [^{14}C]mannitol at similar relative rates as in native membranes. The specific activities of the two reactions, however, were increased about 200-fold relative to the crude extract.

The vectorial transphosphorylation reaction has been reconstituted by incorporation of purified Enzyme II^{mtl} into artificial proteoliposomes (Leonard and Saier, 1983). The reconstituted transport activity was shown to resemble that in whole cells and membrane vesicles. This result shows that vectorial transphosphorylation is catalyzed by the Enzyme II^{mtl} in the absence of the other protein constituents of the PTS. The Enzyme II^{mtl} is therefore the mannitol permease.

Attempts to demonstrate the presence of dimeric Enzyme II^{mtl} using crosslinking reagents with the purified enzyme in detergent micelles or phospholipid bilayers have been largely unsuccessful (Jacobson *et al.*, 1983b). However, several indirect experiments have suggested that the transphosphorylation reaction (both vectorial and nonvectorial) is catalyzed by a dimeric species. The evidence was initially interpreted to suggest that while the transphosphorylation reaction was dependent on dimerization, the phosphoenolpyruvate-dependent phosphorylation of mannitol was catalyzed by monomeric Enzyme II (Saier and Leonard, 1983). It now appears that the dimeric species may catalyze both reactions. The evidence is as follows:

1.   When the two activities were plotted versus enzyme concentration at a constant detergent or phospholipid concentration, a hyperbolic curve was obtained for the phosphoenolpyruvate-dependent reaction, but a sigmoidal response was observed for transphosphorylation. A plot of the rate of transphosphorylation versus the square of the enzyme concentration gave a linear plot. Such a result would be expected if bidirectional transport requires the transient interaction of two subunits while unidirectional transport involves a single subunit. It has since been found, however, that when the phosphoenolpyruvate-dependent reaction is assayed in the presence of mannitol-1-P under the assay conditions employed for transphosphorylation, a plot of reaction rate versus en-

zyme concentration gives a sigmoidal plot. It was therefore concluded that both uni- and bidirectional group translocation of mannitol may depend on transient Enzyme II^{mtl} protomer interactions. Protomer association appears to be promoted by phosphorylation of the Enzyme II, by inclusion of high salt concentrations in the assay solution and by bovine serum albumin. Inclusion of mannitol-1-P apparently had the opposite effect (F. C. Grenier and M. H. Saier, Jr., unpublished results).

2. The pH optimum of the phosphoenolpyruvate-dependent reaction was found to be near 9.0, but that of the transphosphorylation reaction was 6.0. Since the isoelectric point of the enzyme was shown to be near 6 (Jacobson *et al.*, 1983b), this result may be at least partially explained by subunit associations. Minimal electrostatic repulsion at the isoelectric point should maximize subunit interactions, required for the enzyme to catalyze its phosphoryl transfer reactions.

3. A study of the two reaction rates as a function of temperature yielded a linear Arrhenius plot (log v versus $1/T^\circ K$) for temperatures up to 40°C for the phosphoenolpyruvate-dependent reaction, but a break in the transphosphorylation plot was observed at 23°C, and this was followed by downward curvature. This would be expected if high temperatures prevent proper subunit interactions under the conditions of the transphosphorylation assay but not under the conditions of the phosphoenolpyruvate-dependent reaction.

4. Bovine serum albumin, which is known to promote protomer interaction of certain proteins (Glaser and Conrad, 1980), selectively stimulated transphosphorylation while slightly inhibiting the phosphoenolpyruvate-dependent reaction. Other proteins, including Enzyme I, HPr, hemoglobin, and trypsin inhibitor, as well as EDTA, had no effect (Saier and Leonard, 1983). This result was also in accord with expectation assuming that the Enzyme II was predominantly in the monomeric form under the conditions of the transphosphorylation assay, but present as the dimer under the conditions of the phosphoenolpyruvate-dependent assay.

Taken together, these indirect but independent lines of evidence argue that a dimeric or oligomeric species of the Enzyme II^{mtl} polypeptide chain catalyzes sugar group translocation. Under optimal conditions of assay of the phosphoenolpyruvate-dependent reaction, the dimeric form of the enzyme may predominate, and consequently, hyperbolic kinetics are observed when reaction rate is plotted versus enzyme concentration. The conditions of assay of transphosphorylation, particularly the presence of mannitol-1-P, induce dissociation of the dimer to the monomeric species. Consequently, sigmoidal kinetics are observed when reaction rate is plotted against enzyme concentration. Moreover, reaction rate is markedly dependent on temperature and the net charge of the protein, and bovine serum albumin greatly stimulates the reaction. Direct physicochemical approaches will be required to substantiate these postulates. It is worth noting, however, that this interpretation explains how a phosphorylated protein possessing a single sugar substrate binding site can exhibit Bi-Bi Sequential kinetics. If the proposed mechanism is correct, then a half-of-sites mechanism (flip flop or alternating sites mechanism) may be operative (Huang *et al.*, 1982).

Recently the Enzyme II^{glc} has been shown to exist in detergent micelles as a homodimer. Equilibrium centrifugation analyses and chemical cross-linking studies led to the same conclusion (B. Erni, unpublished results). While the molecular weight of the monomer is near 50,000, that of the dimer was reported to be 106,000 ± 5000. Immunoprecipitation of Enzyme II^{glc} yielded a complex of the Enzyme II^{glc} with the Enzyme III^{glc}. Two Enzyme II^{glc} protomers were suggested to be present in association with Enzyme III^{glc}. The demonstration of a dimeric structure for the Enzyme II^{glc} lends further support to the proposal that group translocation of a sugar substrate of the PTS depends upon subunit interactions.

The substrate specificity of the mannitol Enzyme II has recently been examined (Jacobson *et al.*, 1984). The enzyme was shown to be highly specific for D-mannitol, but variations in the orientation of the hydroxyl group at C-2, and variations in the nature of the substituent at the C-2 position were allowed. Thus, glucitol, 2-deoxymannitol (2-deoxyglucitol) and 2-amino-2-deoxymannitol were low-affinity substrates. D-Mannonic acid and D-arabinitol (arabitol) were also substrates. D-Mannoheptitol was shown to bind to the active site although phosphorylation of this sugar could not be demonstrated. Less extensive substrate specificity studies for the three hexitol Enzymes II have been published (Lengeler *et al.*, 1981).

In order to understand the structure–function relationships of the Enzyme II^{mtl}, the gene encoding this protein (the *mtlA* gene) was cloned and sequenced (Lee and Saier, 1983a,b). From the gene sequence, the primary translation product was predicted to consist of

637 amino acyl residues (MW 67,893). The predicted amino acid composition, based on the DNA sequence, was in good agreement with the amino acid composition of the purified protein, except that the hydrophobic amino acids were underestimated by the latter technique. Thus, isoleucine, leucine, and valine were substantially underestimated, and the value for phenylalanine was also somewhat low. Underestimation of hydrophobic residues in integral membrane proteins following acid hydrolysis has been reported by several investigators (see discussion in Lee and Saier, 1983b).

The hydrophobic versus hydrophilic (hydropathy) properties of the enzyme were evaluated along its amino acid sequence using the computer program of Kyte and Doolittle (1982). The results are shown in Fig. 3.2. The analysis predicts that the amino terminal half of the protein resides largely within the membrane. It may traverse the membrane seven or more times. The seven bars under the hydropathy plot at the N-terminus of the protein indicate the seven regions which are of sufficient length and hydrophobic character to pass through the membrane as α-helicies. Computer analyses revealed that β-turns probably exist between each of the hydrophobic stretches except for that between stretches 4 and 5. Moreover, β-turns were predicted at residues 200, 220, and 236, between the shorter hydrophobic segments which might transverse the membrane in the β-configuration (Fig. 3.2). These results substantiate the conclusion that the polypeptide chain of the enzyme traverses the membrane several times.

The entire C-terminal half of the enzyme exhibits the properties of a soluble, globular protein and does not appear to be embedded in the membrane. Computer analyses of this half of the protein suggested that it consists primarily of short α-helicies, separated by β-turns. The only β-structure predicted was between residues 370 and 390, where a likely candidate for an active histidyl residue exists (C. A.

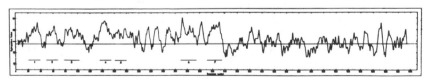

Fig. 3.2. Hydropathic profile of the Enzyme II^{mtl} amino acid sequence. The hydropathy analysis was conducted as described in the legend to Fig. 2.2. As shown, seven hydrophobic regions of 19 residues or greater are indicated by the numbered lines below the graph. [From Lee and Saier (1983b), with permission.]

Lee, R. Doolittle, and M H. Saier, Jr., unpublished observation). Circular dichroism studies may provide some verification for this predicted structure.

The use of antibodies directed against the purified Enzyme II^{mtl} as well as proteolytic enzymes and membrane-impermeable chemical reagents directed against specific amino acyl residues in proteins indicated that the vast majority of the exposed residues and therefore the entire C-terminal half of the protein must be localized to the cytoplasmic surface of the membrane (Jacobson *et al.*, 1983a,b; Saier and Schmidt, 1981). These results also show that Enzyme II^{mtl} is disposed in a single asymmetric orientation in the *E. coli* cytoplasmic membrane.

F. EVOLUTION OF THE PHOSPHOTRANSFERASE
SYSTEM

The possible evolutionary origin of complex phosphotransferase systems such as those found in *E. coli* and *S. typhimurium* has been considered in several earlier articles (Lengeler, 1975a,b, 1977; Lengeler and Steinberger, 1978a,b; Saier, 1977; Saier and Leonard, 1983). As discussed in these articles, the PTS may have evolved from a much simpler system, possibly a phosphoenolpyruvate-dependent sugar kinase which, by virtue of its hydrophobic amino acid composition, became membrane-associated in a fashion which allowed the active site to span the membrane. Thus, it became possible for the sugar substrate to bind to the enzyme surface localized to the external leaflet of the phospholipid bilayer, even though the sugar phosphate product was released into the cytoplasm. In this way, vectorial phosphorylation was accomplished. The anaerobic environment of early biotic earth and the evolution of glycolysis as the first carboydrate catabolic pathway were responsible for the fact that phosphoenolpyruvate (the end product of glycolysis) instead of ATP served as the primary phosphoryl donor for sugar uptake. By virtue of the use of this phosphoryl donor for sugar uptake, the PTS plus the enzymes of glycolysis comprise a "glycolytic cycle" (Fig. 3.1).

Early events leading to a more complex phosphotransferase system may have involved intragenic duplications, gene duplications [possibly resulting from whole operon duplications], or even entire genome duplications (Riley and Anilionis, 1978). Gene segmentation events, giving rise to a multigenic system encoding a number of distinct energy coupling and transport proteins may have followed. The arguments in favor of this supposition have been detailed in the earlier reviews (Saier, 1977; Saier and Leonard, 1983) and will not be discussed further here.

The proposed pathway for the evolution of the sugar-specific enzymes of the PTS is reproduced in Fig. 3.3. The 10 different lines of evidence which led to the postulation of the scheme shown in Fig. 3.3 are summarized in Table 3.3. It is suggested that the primordial Enzyme II exhibited specificity toward fructose and that it was the duplication of the fructose phosphotransferase gene(s) which led to two divergent pathways: the three hexitol phosphotransferases on the one hand, and the four aldohexose phosphotransferases on the other hand. The central role of fructose in glycolysis; the presence of fructose-specific phosphotransferases in numerous gram-negative

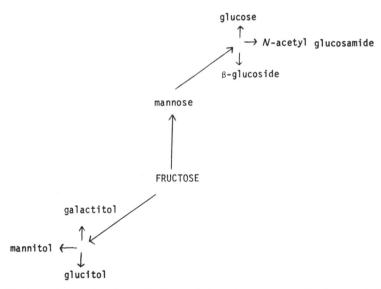

Fig. 3.3. Proposed pathway for the evolution of the sugar-specific Enzyme II–Enzyme III pairs of the phosphotransferase system. This scheme is based on the apparent relatedness of the Enzymes II of the *E. coli* PTS. The scheme indicates directionality in the evolutionary process with the fructose system (middle) as the primordial system giving rise to the hexose-specific systems (top) and the hexitol-specific systems (bottom). [From Saier and Leonard (1983), with permission.]

TABLE 3.3

Evidence for Evolution of the Complex PTS from a Simple Fructose-Specific PTS

1. Fructose is the only sugar which feeds directly into glycolysis
2. Many primitive photosynthetic, N_2-fixing, and heterotrophic bacteria have a fructose-specific PTS
3. Only the fructose regulon has its own HPr
4. There is sequence homology between HPr and an internal region of Enzyme II^{mtl}
5. There are numerous similarities between the Enzyme II^{mtl} and the Enzyme II^{gut}– Enzyme III^{gut} pair
6. All Enzyme II–Enzyme III pairs have similar molecular weights
7. The N-terminal regions of the Enzyme III^{glc} in *E. coli* and the Enzyme III^{lac} in *S. aureus* bind to the Enzymes II
8. HPr and FPr are both phosphorylated on N-1 his
9. All Enzymes III and independent Enzymes II are phosphorylated on N-3 his
10. All Enzymes II may also be phosphorylated on N-1 his

bacteria including primitive photosynthetic (Saier *et al.*, 1971), nitrogen-fixing (K. Basu and S. Ghosh, personal communication), and heterotrophic bacteria; and the presence of a fructose-inducible HPr-like molecule, presumably encoded within the *fru* regulon but lacking in the other sugar-specific operons encoding PTS proteins, all argue in favor of this postulate (Saier and Leonard, 1983).

Recently, new information has become available which appears to substantiate these proposed relationships. As described in the previous section, the mannitol Enzyme II has been purified to homogeneity, and many of its physicochemical properties have been determined. Moreover, the *mtlA* gene encoding this protein has been sequenced, and the primary aminoacyl sequence of the protein was deduced from the gene sequence. While the protein migrates on SDS gels as a protein of molecular weight equal to 60,000, the molecule possesses a hydrophobic amino terminal half, presumably embedded in the membrane, and a hydrophilic C-terminal half, localized to the cytoplasmic surface of the membrane. These and other properties of the enzyme are summarized in Table 3.4.

Recent studies on the glucitol phosphotransferase revealed that this system consists of an Enzyme II^{gut}–Enzyme III^{gut} pair as shown in Fig. 3.4 (Grenier *et al.*, 1985; Sarno *et al.*, 1984). The properties of the Enzyme II^{gut} are strikingly similar to those of the Enzyme II^{mtl} as revealed in Table 3.4. In view of the previously proposed close evolutionary relationships between the *mtl* and *gut* operons, (Lenge-

TABLE 3.4

Comparative Aspects of the Mannitol- and Glucitol-Specific Enzymes of the Phosphotransferase System in *S. typhimurium*

	Enzyme III^{gut}	Enzyme II^{gut}	Enzyme II^{mtl}
Molecular weight (PAGE)	16,000	45,000	60,000
K_m (gut)			
transport		$12\mu M$	$2500\ \mu M$
phosphorylation		$44\mu M$	$400\ \mu M$
K_m (mtl)			
transport		$3300\ \mu M$	$0.4\ \mu M$
phosphorylation		$60\ \mu M$	$3\ \mu M$
Induction by glucitol	+	+	±
Induction by mannitol	−	−	+
Solubilization by DOC/NaCl		+	+
Purification by hexylagarose		+	+
NEM sensitivity	−	+	+
DEPC sensitivity	+	+	+
Anti-Enzyme II^{mtl} antibody	±	−	+
Anti-Enzyme III^{gut} antibody	+	−	−

ler, 1975a,b, 1977; Lengeler and Steinberger, 1978a,b; Saier and Leonard, 1983) and the fact that only the A and D structural genes had been detected by earlier genetic analyses (Lengeler, 1975a), the discovery of an essential Enzyme III^{gut} was surprising. A possible explanation for this difference resulted from the purification of the Enzyme II^{gut} which revealed that while the Enzyme III^{gut} migrated in SDS gels with a molecular weight of 16,000, the apparent molecular

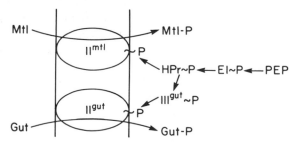

Fig. 3.4. Sequential phosphorylation reactions of the phosphoryl transfer chains for mannitol (Mtl) and glucitol (Gut). The designations of the PTS proteins are as shown in Fig. 3.1.

weight of the Enzyme II^{gut} was about 45,000. Thus, the sum of the molecular weights of the Enzyme II^{gut}–Enzyme III^{gut} pair is approximately equal to that of the Enzyme II^{mtl} (Table 3.4). The two *gut* proteins might have arisen by the introduction of a nonsense mutation within a single primordial gene which has remained intact in the *mtl* operon. Consistent with this conclusion is the demonstration that the order of genes within the *gut* operon is probably *gutCABD*, where A codes for the Enzyme II^{gut}, B codes for the Enzyme III^{gut}, D codes for glucitol-6-P dehydrogenase, and C is a regulatory gene. The position of the B gene in the operon would suggest that the Enzyme III^{gut} is structurally related to the hydrophilic carboxy-terminus of the Enzyme II^{mtl} (Sarno *et al.*, 1984).

A separate line of evidence leading to the conclusion that the Enzyme III^{gut} is structurally related to the C-terminus of the Enzyme II^{mtl} resulted from studies with antibodies directed against the purified Enzyme II^{mtl}. This antibody was found to inhibit Enzyme II^{mtl} only from the cytoplasmic side of the membrane (Jacobson *et al.*, 1983b). It was also found to inhibit glucitol phosphorylation catalyzed by the Enzyme II^{gut}–Enzyme III^{gut} pair, but surprisingly, inhibition could be overcome by increasing the concentration of Enzyme III^{gut} rather than that of Enzyme II^{gut} (F. C. Grenier, unpublished observation). This result suggested that the principal inhibitory antibodies were directed against epitopes on the Enzyme II^{mtl} which were similar to sites on the Enzyme III^{gut}.

If the *mtl* and *gut* operons arose from a common ancestral operon, the genes encoding the two dehydrogenases might also share sequence homology which would be reflected in the properties of these enzymes. The mannitol-1-P and glucitol-6-P dehydrogenases have been purified to near homogeneity, and their properties have been compared (Table 3.5; M. J. Novotny, J. Reizer, F. Esche, and M. H. Saier, Jr., unpublished results). The two proteins have quite different subunit molecular weights as determined by sodium dodecyl sulfate gel electrophoresis (40,000 versus 28,000) and their oligomeric structures differ. While the former enzyme is a monomer in solution, the latter enzyme behaves like a tetramer. The two enzymes show absolute specificities for their hexitol-phosphate substrates and exhibit normal Michaelis–Menten kinetics with absolute binding constants (K_m values) of 0.8 and 0.2 mM for mannitol-1-P and NAD$^+$, respectively, for the mannitol-1-P dehydrogenase and of 3 and 0.2 mM for glucitol-6-P and NAD$^+$, respectively, for the gluci-

TABLE 3.5

Comparative Aspects of the Mannitol-1-P and Glucitol-6-P
Dehydrogenases of *E. coli*

Property	Mannitol-P dehydrogenase	Glucitol-P dehydrogenase
Subunit molecular weight	40,000	28,000
Oligomeric structure	Monomer	Tetramer
Substrate specificity	Absolute for mannitol-1-P	Absolute for glucitol-6-P
NEM sensitivity	−	±
DEPC sensitivity	−	+

tol-6-P dehydrogenase. Of these two enzymes only the glucitol-6-P dehydrogenase was inactivated by *N*-ethylmaleimide (NEM) and diethylpyrocarbonate (DEPC), reagents which, respectively, derivatize sulfhydryl and histidyl residues in proteins (Table 3.5). Amino acid compositions and N-terminal aminoacyl sequences were also very different. These results show that the two dehydrogenases differ markedly with respect to many of their properties. The data do not provide evidence for a common ancestral origin for these two genes.

As mentioned above, the molecular weight of Enzyme II^{mtl} is about equal to those of the Enzyme II^{gut} and Enzyme III^{gut} combined. Comparisons of the molecular weights of the Enzymes II and III for different sugars (Table 3.6) led to a startling observation: The molecular weights of the sums of the four known Enzyme II–Enzyme III pairs are equal to about 68,000, the value of the mannitol Enzyme II molecular weight. Thus, the respective values reported for the glucose pair are 45,000 and 21,000 (sum = 68,000); those for the mannose pair are 25,000 and 35,000 (sum = 60,000) or 36,000 and 33,000 (sum = 69,000; see Table 3.6); and those for the lactose pair in *Staphylococcus aureus* are 55,000 and 12,000 (sum 67,000). The *N*-acetylglucosamine Enzyme II, which functions without the participation of an Enzyme III has a molecular weight of 65,000 (E. B. Waygood, personal communication). These remarkable observations suggest a common evolutionary origin for these phosphotransferase proteins.

If the Enzyme II–Enzyme III pairs resulted from segmentation of a primordial Enzyme II structural gene, then the N-terminus of the

TABLE 3.6

Molecular Weight Comparisons for Various Enzyme II–Enzyme III
Pairs of the PTS

Proteins[a] specific for	Molecular Weight[b] $\times$ 10^{-3} of		
	Enzyme II	Enzyme III	Sum
Mannitol	60	None	60
	(68)		(68)
Glucitol	45	16	61
Glucose	45	21	66
Mannose	25	35	60
	(36)	(33)	(69)
N-Acetylglucosamine	65	None	65
Fructose	—	40	—
Lactose	55	12	67

[a] All proteins are from *E. coli* with the exception of the lactose
Enzyme II–Enzyme III pair which is from *Staphylococcus aureus*.

[b] Molecular weights were usually estimated by SDS-polyacrylamide gel electrophoresis. The value in parenthesis for the mannitol
Enzyme II was deduced from the *mtlA* gene sequence, while the values for the mannose Enzyme II–Enzyme III pair are those obtained by
B. Erni (unpublished results). Values in parentheses were reported by
Kundig and Roseman (1971) and E. B. Waygood (unpublished results)
for the Enzyme II[man] and Enzyme III[man], respectively.

Enzyme III would be expected to have been continuous with the C-
terminus of the Enzyme II in the unsegmented, primordial protein.
Assuming this to be true, the N-terminus of an Enzyme III might be
expected to be contiguous with the C-terminus of its Enzyme II in
the Enzyme II–Enzyme III complex. In other words, the N-terminus of an Enzyme III should function in binding to its homologous
Enzyme II in the membrane, while the C-terminal region might be
expected to interact with HPr. Recent preliminary evidence appears
to substantiate this prediction. First, a proteolytically clipped Enzyme III[glc], in which the seven amino terminal aminoacyl residues
had been removed, was fully active in phosphoryl transfer from
phospho-HPr, but deficient in its activity as a phosphoryl donor for
methyl α-glucoside phosphorylation in the presence of Enzyme II[glc]
(Meadow *et al.*, 1982a). Second, derivatization of the N-terminal
glycyl residue in the *E. coli* Enzyme III[glc] with fluorescein gave a
modified protein which was phosphorylated normally by phospho-

HPr, but was only one-tenth as active as native phospho-Enzyme III^{glc} in phosphoryl transfer to Enzyme II^{glc} (Jablonski *et al.*, 1983). Third, a tryptic fragment of the Enzyme III^{lac} of *Staphylococcus aureus*, consisting of the first 38 residues at the amino terminal part of the protein, specifically inhibited the interaction between Enzyme II^{lac} and Enzyme III^{lac} (Deutscher *et al.*, 1982). Finally, a mutant of *S. aureus* has been isolated in which the Enzyme III^{lac} showed a single amino acid substitution: the glycyl residue in position 18 was replaced by a glutamyl residue. This defect did not prevent phosphoryl transfer from phospho-HPr but abolished that to Enzyme II^{lac} (H. M. Sobek, K. Stüber, K. Beyreuther, W. Hengstenberg, and J. Deutscher, unpublished results). These observations, taken together suggest that the N-termini of both the Enzyme III^{glc} of *E. coli* and the Enzyme III^{lac} of *S. aureus* function to bind the proteins to the Enzymes II in the cytoplasmic surface of the membrane.

An antibody preparation directed against the mannitol Enzyme II was tested for its inhibitory effect on other Enzymes II of the PTS. The relative inhibitory effects of low antibody concentrations on the activities of the different phosphotransferases were as follows (Jacobson *et al.*, 1983b):

$$mtl > gut > fru > man > glc = nag$$

The order of relatedness predicted from this result was as expected based on the evolutionary scheme shown in Fig. 3.3.

The proposed evolutionary scheme suggests that the different aldohexose-specific Enzymes II diverged relatively recently from one another. In connection with this suggestion, two additional Enzymes II have been characterized in gram-negative bacteria which exhibit specificity for glucopyranosyl residues. One of these, a sucrose-specific Enzyme II, is plasmid encoded in *S. typhimurium* and chromosomally encoded in *Klebsiella pneumoniae*. It phosphorylates the glucosyl moiety of sucrose (Lengeler *et al.*, 1982). The other Enzyme II from *Vibrio parahaemolyticus* is specific for trehalose. It phosphorylates one of the two symmetrical glucosyl residues in trehalose (Kubota *et al.*, 1979). In view of their substrate specificities, both of these Enzymes II might be expected to be related to the glucose Enzyme II. In fact, the sucrose Enzyme II, and possibly the trehalose Enzyme II as well, has been reported to depend on the glucose Enzyme III for activity (Lengeler *et al.*, 1982). It is therefore reasonable to propose that these Enzymes II diverged from the glucose specific Enzyme II during recent evolutionary history with the retention of the same Enzyme III. Sequence analyses will be required to establish this proposed relationship.

If the different PTS enzymes did, in fact, evolve from a common ancestral protein and if the *fru* regulon, which presumably encodes Enzyme II^{fru}, Enzyme III^{fru}, and a fructose-inducible HPr-like protein (FPr) is related to the primordial gene, then the sequence of one PTS protein might be related to parts of the sequences of other PTS proteins. We have surveyed for possible sequence homology employing computer techniques and have discovered significant sequence homology between a part of *Salmonella* HPr and a part of the *E. coli* Enzyme II^{mtl} (M. H. Saier, Jr. and R. Doolittle, unpublished observations). The significance of the sequence homology observed between HPr and Enzyme II^{mtl} is emphasized by the fact that no other prokaryotic, eukaryotic, or viral protein screened showed significant homology with either HPr or Enzyme II^{mtl}. Part of this sequence homology is illustrated in Fig. 3.5. Identical residues are boxed with solid lines while near identical residues (resulting from a single base substitution in the triplet code) are boxed with dashed lines. As can be seen, the degree of homology is striking. In

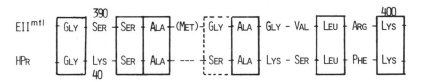

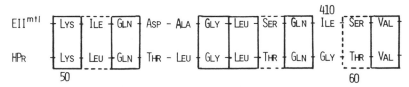

Fig. 3.5. Partial sequence homology between portions of HPr from *S. typhimurium* and the Enzyme II^{mtl} of *E. coli*. Those residues which are identical are boxed with solid lines. Those which differ with respect to a single base substitution are boxed with dashed lines. Fifty percent of the residues within this region are identities, and an additional 17% are near identities resulting from a single base substitution. The region of partial homology between these two proteins is more extensive than shown accounting for a total length of about 50 aminoacyl residues in each of the two proteins.

the most homologous region, the number of identities plus near-identities is about two-thirds of the total amino acids. It is worthy of note that the region of partial homology is much more extensive than that shown in Fig. 3.5. Interestingly, in the mannitol Enzyme II, a 15 aminoacyl insert of nonhomology is found within an extensive region of partial homology. This 15 aminoacyl residue segment carries a histidyl residue which is a prime candidate for an active site residue. The partially homologous segment of HPr is immediately adjacent to and to the left of the active histidyl residue in HPr (residue no. 15). These provocative observations provide preliminary evidence for an evolutionary scheme in which a primordial operon encoding an HPr-like protein (as is suggested for the *fru* regulon) served as a precursor gene for the evolution of genes encoding sugar-specific proteins of the complex PTS found in enteric bacteria. Table 3.3 summarizes the different lines of evidence which support the evolutionary scheme depicted in Fig. 3.3. More extensive sequence analyses, particularly of the *gut* and the *fru* operons, should allow more definitive conclusions with respect to the ancestry of the PTS proteins.

4

Mechanisms of Inducer Exclusion

The synthesis of carbohydrate catabolic enzyme systems in gram-negative bacteria is generally regulated by a dual mechanism dependent on two small cytoplasmic molecules: inducer and cyclic AMP (Fig. 4.1). Each of these molecules acts through a macromolecular effector. The effector protein mediating the action of the inducer is a transcriptional repressor such as the lactose repressor (product of the *lacI* gene) or the glycerol repressor (product of the *glpR* gene), or it is a transcriptional activator such as the maltose activator (product of the *malT* gene). Repressor proteins exert negative control over target operon expression by binding to the operator and preventing transcription by RNA polymerase. An increase in the

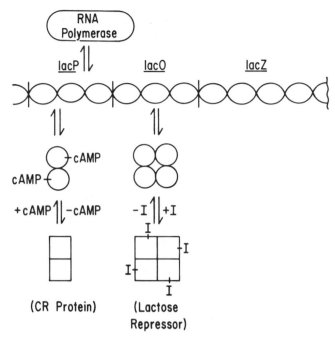

Fig. 4.1. Proposed scheme for the regulation of the transcription of the lactose operon in *E. coli*. The RNA polymerase binds to a specific site in the promoter region of the lactose operon (*lacP*). It can transcribe the DNA sequence of the structural genes of the operon into messenger RNA only if the cyclic AMP–CR protein complex is bound to the promoter region of the operon (positive control) and if the repressor is *not* bound to the operator region (*lacO*) (negative control). The scheme illustrates dual control of transcription by two small cytoplasmic molecules, cyclic AMP, and the inducer (I). The first structural gene in the *lac* operon, *lacZ*, codes for β-galactosidase. [From Saier and Stiles (1975), with permission.]

cytoplasmic inducer concentration results in the binding of inducer to the repressor, causing it to assume a conformation with low affinity for the operator. Consequently, the inducer promotes transcription by causing dissociation of the repressor from the control region of the operon, relieving a negative regulatory mechanism. An activator protein such as the product of the *malT* gene, on the other hand, binds the inducer (maltose), and the inducer–activator complex promotes transcription. Maltose causes the activator to assume a conformation which possesses high affinity for the operator, and binding of the inducer-activator complex to the DNA promotes operon transcription by a positive control mechanism. Cyclic AMP exerts its effect on carbohydrate catabolic gene expression by binding to the bivalent cyclic AMP receptor protein (CRP), inducing a conformation which binds to the promoter regions of the various operons under cyclic AMP control, and thereby promoting transcription by RNA polymerase (Fig. 4.1). While much information concerning the modes of action of repressors, activators and the cyclic AMP receptor protein is emerging (Bourgeois and Pfahl, 1976; O'Neill *et al.*, 1981; McKay and Steitz, 1981; Steitz *et al.*, 1982; Ohlendorf *et al.*, 1982; Ullman and Danchin, 1983), this chapter will concentrate on the mechanisms by which intracellular concentrations of inducer are controlled.

In 1967, McGinnis and Paigen measured the rates of utilization of ^{14}C-labeled carbohydrates in the presence and absence of a second nonradioactive carbohydrate. They demonstrated that glucose caused an immediate and reversible inhibition of ^{14}C-labeled sugar utilization. In a subsequent publication, these same investigators proposed that glucose exerted its inhibitory effects on carbohydrate uptake systems (McGinnis and Paigen, 1973). However, the mechanisms of inhibition were not considered.

As a result of studies conducted during the past decade, five regulatory mechanisms, which appear to account for the control of carbohydrate uptake via various permease systems, have been proposed. These include (i) inhibition due to competition between two sugars for sugar binding to the substrate recognition site of a permease protein, (ii) PTS-mediated regulation involving the allosteric inhibition of permease activity by a protein of the PTS, (iii) inhibition of carbohydrate uptake by intracellular sugar phosphates, (iv) chemiosmotic control of transport activity due to responsiveness of the permease system to the membrane potential and/or the intracellular pH, and (v) inhibition of the uptake of one PTS sugar substrate by another sugar substrate of this system due to competition of the corresponding Enzyme II complexes for the common phosphoryl protein, phospho-HPr. These mechanisms are illustrated schematically in Fig. 4.2.

In this and the subsequent chapter, it will be seen that usually, the transport systems responsible for inducer uptake as well as adenylate cyclase, the cyclic AMP biosynthetic enzyme, are the primary targets of regulation. Both are regulated by mechanisms which allow the bacteria to sense and respond to the availability of various en-

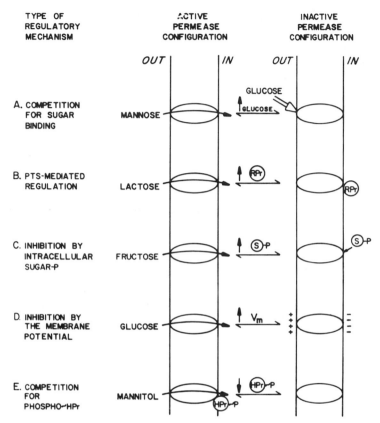

Fig. 4.2. Proposed carbohydrate transport regulatory mechanisms in bacteria. Representative examples occur in *E. coli*. Abbreviations: HPr, histidine-containing phosphocarrier protein of the PTS; PTS, phosphoenolpyruvate : sugar phosphotransferase system; RPr, regulatory protein (glucose Enzyme III [Enzyme III^{glc}] of the PTS); sugar-P, sugar phosphate; V_m, membrane potential or possibly the proton electrochemical gradient; S, substrate. [From Dills *et al.* (1980), with permission.]

ergy sources as discussed more extensively in Chapter 5. This fact largely accounts for the regulation of transcriptional initiation. of carbohydrate catabolic enzyme systems. As a result of recent studies, it has been shown that each of the major mechanisms controlling transport function is operative in both gram-negative and gram-positive bacteria. These regulatory mechanisms controlling inducer uptake are the subjects of this chapter.

A. REGULATION OF TRANSPORT BY THE MEMBRANE POTENTIAL

It has long been recognized that chemiosmotic energy influences the activities of some transport systems in a positive sense while exerting negative control over others. For example, the citrate and phosphoglycerate permeases in *Salmonella typhimurium* require a substantial membrane potential, even for downhill uptake of their substrates (Saier, 1979). One possible explanation for this observation is that the dipole moments of the permease proteins interact with the membrane potential in a way which influences the conformation of the protein. This possibility is illustrated in Figure 4.3. The permeases (Enzymes II) of the phosphotransferase system are inhibited by chemiosmotic energy (for reviews of the early literature, see Saier and Moczydlowski, 1978; Saier, 1979). The latter observation was confirmed and extended by Reider *et al.* in 1979. Uptake of methyl α-glucoside, a nonmetabolizable substrate of the glucose phosphotransferase, was studied in *E. coli* cells and membrane vesicles. The use of *E. coli* cells defective for the proton-translocating ATPase, due to mutations in the *uncA* and *uncB* genes, prevented the rapid interconversion of chemical and chemiosmotic energy and facilitated interpretation of the results.

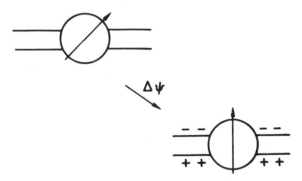

Fig. 4.3. Theoretical effect of an imposed membrane potential on the dipole moment of a transmembrane protein. If the electrical dipole of the protein is nonperpendicular to the plane of the membrane, imposition of an electrical potential across the membrane will tend to convert the protein to a conformation with its dipole more perpendicular to the plane of the membrane. If these two conformations differ in biological transport activity, the permease or channel protein will be regulated by the potential. [From Saier and Jacobson (1984), with permission.]

When wild-type cells and the *uncA uncB* double mutant were studied under aerobic conditions, inhibitors of electron transport and proton-conducting uncouplers greatly stimulated methyl α-glucoside uptake. Under anaerobic conditions uptake rates were comparable to the aerobic rates in the presence of an uncoupler, and the addition of uncouplers did not stimulate uptake. Anaerobic cells were shown to exhibit a low state of energization as shown by the fluorescence yield in the presence of 1-anilino-8-naphthalene sulfonate (ANS).

In membrane vesicles, D-lactate an efficient electron donor, and cyanide, an inhibitor of electron flow, were found to exert opposite effects on methyl α-glucoside uptake. Oxamate, a potent inhibitor of D-lactate dehydrogenase, and cyanide, reversed the inhibitory effect of lactate. Lactate was shown to energize proline uptake under the same conditions, while oxamate and cyanide prevented energization.

When a variety of potential electron donors were examined for their effects on sugar uptake, a correlation was noted. Those donors which delivered electrons most efficiently to the electron transport chain were most effective in inhibiting methyl α-glucoside uptake. Moreover, qualitatively similar effects were observed when the radioactive sugar substrate was glucose, fructose, or mannose instead of methyl α-glucoside. An intact respiratory chain was presumed to be essential for the inhibitory effect since a heme-deficient mutant showed high rates of methyl α-glucoside accumulation with no response to D-lactate in vesicles. It was concluded that the activity of the phosphotransferase system is negatively responsive to chemiosmotic energy (Reider *et al.*, 1979). Neither the mechanism of inhibition nor the target enzyme of the PTS which is subject to inhibition was defined in these studies.

In a subsequent paper Robillard and Konings (1981) attempted to provide a mechanistic interpretation to these results. They showed that sugar uptake via the PTS is strongly inhibited when the redox potential exceeds 300 mV (negative inside), regardless of whether the membrane potential is generated by substrate oxidation or by direct addition of an oxidizing agent. Dithiothreitol prevented the inhibitory effect suggesting that the reversible oxidation of a pair of sulfhydryl groups might be involved. While inhibition by D-lactate, NADH, or reduced phenazine methosulfate could be reversed by proton-conducting uncouplers, the inhibitory effects of oxidizing agents were not. It was concluded that the membrane potential, generated in response to electron donors, regulates the glucose phosphotransferase by shifting the midpoint potential of the Enzyme II–associated redox transition to more negative values.

According to this proposal, illustrated in Fig. 4.4, an oxidizing agent or the membrane potential converts the reduced (dithiol) form of the glucose Enzyme II (Enzyme II^{glc}) to an oxidized (disulfide) form. Similar effects were observed for sugar uptake in right-side-out vesicles and for sugar phosphorylation in inverted vesicles. Based on kinetic studies, which revealed that inhibition was strong at low methyl α-glucoside concentration (5 μM) but virtually nonexistent at high substrate concentration (5 mM), it was proposed that the reduced (dithiol) form of the Enzyme II possesses high affinity for substrate while the oxidized (disulfide) form exhibits low affinity (Fig. 4.4). The difference in apparent affinity of each of these two forms was reported to be about 100-fold.

An alternative explanation for these results suggests that only the reduced form of the enzyme possesses catalytic activity and binds substrate. The oxidized form is essentially inactive. According to this interpretation, the lack of inhibition at high sugar substrate concentration is attributed to a shift of the equilibrium towards the

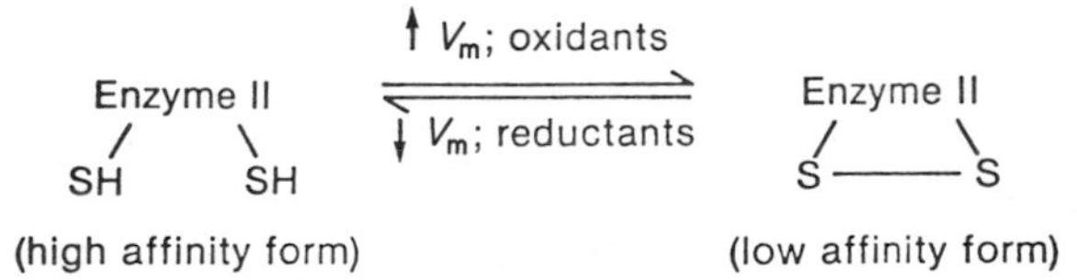

Fig. 4.4. Schematic depiction of the postulated reversible oxidation of the Enzymes II of the bacterial phosphotransferase system. The process is thought to involve the interconversion of a dithiol (reduced) form of the enzyme and a disulfide (oxidized) form of the enzyme. Oxidants such as ferricyanide or oxidized glutathione convert the enzyme to the disulfide form while disulfide reductants such as dithiothreitol or reduced glutathione reverse enzyme oxidation, generating the dithiol form of the enzyme. An increase in the magnitude of the membrane potential (V_m) (negative inside) is presumed to promote enzyme oxidation. Sugar binding is thought to promote enzyme reduction because the reduced form of the enzyme possesses about 100-fold higher affinity for the substrate than the oxidized form. Conversely, enzyme phosphorylation may enhance the extent of enzyme oxidation. This model provides a possible molecular explanation for the regulation of permease function by the membrane potential as suggested by Robillard and Konings (1981).

reduced form of the enzyme, promoted by preferential (or exclusive) binding of sugar to that form.

One criticism of these studies is that they were conducted with crude membrane preparations which contain multiple Enzymes II with differing affinities for hexoses. Additionally the presence of an oxidized form of the Enzyme II with inherent, albeit low, affinity and activity has not been clearly demonstrated. In view of these considerations, the mechanism by which oxidizing agents regulate the activity of the purified mannitol Enzyme II from *E. coli* was studied (F. C. Grenier, E. B. Waygood, and M. H. Saier, Jr., unpublished results). Transphosphorylation experiments verified that oxidizing agents including potassium ferricyanide directly inhibit mannitol Enzyme II activity. Phosphorylation of Enzyme II^{mtl} with Enzyme I, HPr, and phosphoenolpyruvate partially protected the enzyme from ferricyanide inhibition, and the enzyme was even less sensitive to inhibition during catalytic turnover. However, preincubation of the unphosphorylated enzyme with ferricyanide quantitatively but reversibly inactivated it. Activity was not restored when high substrate concentrations were employed during assay. Similar observations were made when the Enzyme II^{mtl} in crude membrane preparations was studied instead of the purified enzyme. The results were therefore inconsistent with a regulatory mechanism in which sulfhydryl oxidation influences the affinity of the enzyme for its sugar substrate. Instead, it was concluded that the oxidized enzyme is inactive. It is not known if these results conflict with those of Robillard and Konings (1981) or if the glucose and mannitol Enzymes II differ in their responses to oxidizing agents.

Konings and Robillard (1982) have extended their observations to transport systems other than the phosphotransferase system Enzymes II. It was proposed that substrate–proton symporters such as the lactose and proline permeases in *E. coli* are regulated by changing the redox state of dithiols within these carriers. In support of this notion, they showed that lipophilic oxidizing agents such as plumbagin, phenazine methosulfate, menadione, and 1,2-naphthoquinone inhibited lactose uptake and that dithiothreitol reversed this inhibitory effect. Phenylarsine oxide, a dithiol-specific reagent, also inhibited lactose uptake, and moreover, the lipophilic oxidizing agents which inhibited lactose uptake protected against irreversible inhibition by *N*-ethylmaleimide. As for the glucose Enzyme II of the PTS, it was suggested that oxidation of a dithiol to a disulfide increases the K_m of the carrier for lactose by 100-fold, from 0.2 to 20 mM. This reversible oxidation was presumed to be effected either by the dissipation of a transmembrane electrical potential (negative inside) or by the addition of a lipophilic oxidizing agent. Facilitated diffusion of lactose presumably reflects the activity of the oxidized carrier only. The requirement for a lipophilic oxidizing agent, to be contrasted with the situation for the Enzymes II where hydrophilic oxidizing agents could be used, might reflect a membranous localization of the essential sulfhydryl group(s) in the lactose carrier. The evaluation of these postulates may be important with respect to the mechanisms by which transport systems are regulated by chemiosmotic energy.

B. REGULATION OF TRANSPORT BY INTRACELLULAR SUGAR PHOSPHATES

Possibly the most general mechanism for regulating the uptake of carbohydrates involves inhibition by intracellular metabolites (Fig. 4.5). While the nature of the inhibitory metabolites has not been fully elucidated, it is at least clear that nonmetabolizable intracellular sugar phosphates inhibit uptake of polyols and hexoses in both gram-negative and gram-positive bacteria and in eukaryotes including yeasts and certain animal cells maintained in culture (Mitchell *et al.,* 1982; Kornberg, 1973; Saier and Simoni, 1976). This type of inhibition is characterized by a lag period for inhibition of uptake by the bacterial suspension containing the radioactive substrate upon addition of a nonmetabolizable sugar which can accumulate in the cell as its phosphate ester. An example of this behavior is shown in Fig. 4.6 (Saier and Simoni, 1976). As shown in Fig. 4.6A, glycerol uptake into *Staphylococcus aureus* cells is strongly inhibited when isopropyl-β-thiogalactoside, a substrate of the lactose PTS in this organism, is added 15 min before addition of the radioactive glycerol. By contrast, when the radioactive glycerol and the inhibitory sugar are added simultaneously, the initial rate of glycerol uptake is equal to the uninhibited rate, and only after several minutes is the inhibitory effect of the galactoside apparent. As shown in Fig. 4.6B,

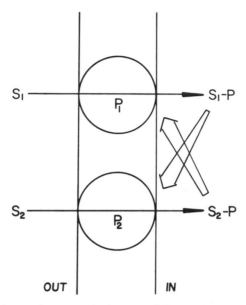

Fig. 4.5. Inhibition of sugar uptake by intracellular metabolites. The figure depicts two permeases (Enzymes II of the phosphotransferase system), P_1 and P_2, which transport and phosphorylate their respective sugar substrates, S_1 and S_2. The intracellular sugar phosphates as well as other metabolites derived from the sugar phosphates inhibit the activities of the permeases by unspecified mechanisms.

uptake of the galactoside is a slow process, and the degree of inhibition can be related solely to the intracellular concentration of the galactoside-phosphate, not to the rate of galactoside uptake (Saier and Simoni, 1976). The kinetics of the inhibitory effect distinguish this type of inhibition from PTS-mediated regulation (Section C) or an energy competition–type mechanism (Section D).

Strong inhibition of [^{14}C]mannitol uptake is observed in *E. coli* or *Salmonella typhimurium* when 2-deoxyglucose is added to the bacterial suspension. The identification of a sugar phosphate binding site on the cytoplasmic surface of the Enzymes II of the PTS (Saier *et al.*, 1977; Rephaeli and Saier, 1980a) led to the possibility that inhibition by intracellular sugar phosphate might result from the direct binding of the inhibitory sugar phosphate to the sugar phosphate binding site of the Enzyme II. To test this possibility, the

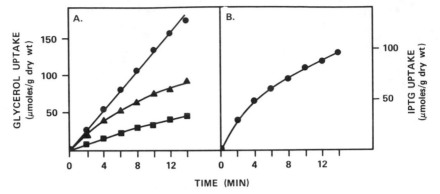

Fig. 4.6. Time courses for the uptake of [¹⁴C]glycerol and [¹⁴C]isopropyl β-thioga-lactoside. *Staphylococcus aureus* strain C22 cells were prepared for the uptake experiments as described in Saier and Simoni, (1976). In experiment A, the initial concentration of [¹⁴C]glycerol was 0.2 mM. Uptake was measured at 28°C in the absence of inhibiting sugar (●), in the presence of 0.67 mM isopropyl β-thiogalactoside, added to the cell suspension 15 min before initiation of glycerol uptake (■), or in the presence of 0.67 mM isopropyl β-thiogalactoside, added simultaneously with [¹⁴C]glycerol (▲). In experiment B, uptake of [¹⁴C]isopropyl β-thiogalactoside was followed in the presence of 0.2 mM nonradioactive glycerol added simultaneously with the radioactive substrate. Uptake is expressed in micromoles of solute accumulated per gram of dry weight of cells. [From Saier and Simoni, (1976), with permission.]

phosphoenolpyruvate-dependent phosphorylating activity of the purified mannitol Enzyme II was studied in the presence of varying amounts of the energy coupling proteins, Enzyme I and HPr, and in the presence and absence of a variety of sugar phosphates including glucose-6-P, fructose-6-P and fructose-1,6-diP. While the concentration of [¹⁴C]mannitol was in the micromolar range, inhibitor concentrations of up to 20 mM were studied (Jacobson *et al.*, 1983b). In no case was the inhibitory effect greater than 30%, and the monophosphate esters were only slightly inhibitory. This result suggested either that the purified Enzyme II differed from the enzyme in intact cells with respect to its sensitivity to inhibition or that inhibition was indirect. The recent discovery of metabolite-activated protein kinases which influence carbohydrate transport and metabolism (Deutscher and Saier, 1983; Wang and Koshland, 1982) leads to the possibility that this type of inhibition is mediated by an ATP-dependent phosphorylation mechanism.

C.   PTS-MEDIATED REGULATION OF CARBOHYDRATE UPTAKE

i.   Early Studies

As discussed in greater detail elsewhere (Dills *et al.*, 1980; Peterkofsky and Gazdar, 1975; Postma and Roseman, 1976; Rosen, 1978; Saier, 1977), the PTS is thought to function catalytically to regulate the uptake of several carbohydrates. In *E. coli* and *S. typhimurium,* these carbohydrates include glycerol, lactose, melibiose, and maltose. The PTS also regulates the activity of adenylate cyclase, the cyclic AMP biosynthetic enzyme, by an apparently coordinate mechanism (Saier and Feucht, 1975). One possible mechanism first proposed in 1975 by Saier and Feucht and subsequently suggested by others (Postma and Roseman, 1976; Gonzales and Peterkofsky, 1977; Peterkofsy and Gazdar, 1978) is illustrated in Fig. 4.7 and discussed below. Although some evidence has been interpreted to suggest that the regulatory role of the PTS is indirect (Yang *et al.*, 1979), other results indicate a direct, catalytic role (Castro *et al.*, 1976; Peterkofsky and Gazdar, 1978; Postma and Roseman, 1976; Saier and Feucht, 1975). The early genetic studies, together with biochemical and physiological analyses (Saier, 1977) led to the following observations:

1.   Reduced cellular activities of either Enzyme I or HPr render target permeases hypersensitive to inhibition by any extracellular sugar substrate of the PTS.
2.   Inhibition by a particular sugar substrate of the PTS requires that the Enzyme II, which recognizes, transports, and phosphorylates that sugar, is catalytically active.
3.   Mutation of a gene (designated *crrA*, which maps adjacent to the *pts* operon) renders all sensitive permeases insensitive to PTS-mediated regulation by all sugar substrates of the PTS although other regulatory mechanisms may still be operative. The *crrA* mutants exhibit permease activities that are comparable to maximal activities observed in the wild type strain. *crrA* mutants exhibit low activity of a particular PTS protein, the glucose Enzyme III, thus implicating this protein in the regulatory process.
4.   Regulatory mutations which map within or very near the

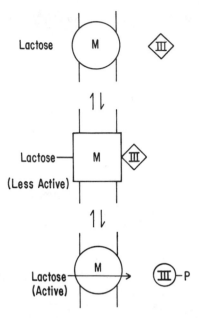

Fig. 4.7. Proposed mechanism of the regulatory interaction between Enzyme III^{glc} and the lactose permease. M and III indicate lactose permease (M protein) and Enzyme III^{glc}, respectively. Lactose and Enzyme III^{glc} are shown to bind to the lactose permease on opposite sides (outer surface for lactose and inner surface for Enzyme III^{glc}) of the plasma membrane (the area between the two vertical lines). Phosphorylation of Enzyme III^{glc} causes it to dissociate from the allosteric regulatory site of the lactose permease. [From Osumi and Saier (1982a), with permission.]

genes that code for the individual target permease proteins abolish the PTS-mediated control of that permease without altering the regulatory constraints imposed on other target permeases (Saier *et al.*, 1978a).

ii. Desensitization

Under appropriate physiological conditions, (when cellular inducer and cyclic AMP levels are high so that the non-PTS permeases are induced to high levels) these target permeases become "desensitized" (insensitive or less sensitive) to PTS-mediated regulation

(Saier *et al.*, 1982b). A detailed discussion of this phenomenon can be found in Dills *et al.* (1980).

The desensitization phenomenon described above has been independently observed by Postma and co-workers (Nelson *et al.*, 1982; Scholte *et al.*, 1981, 1982) and by Mitchell *et al.* (1982) but not by Fraser and Yamazaki (1983). Quantitative measurements of the former investigators led to the conclusion that the degree of inhibition of a given permease is related not only to the level of free (nonphosphorylated) Enzyme III^{glc}, but also to the level of the target permease. Synthesis of high levels of the permease (so that the number of permease proteins exceeds the number of Enzyme III^{glc} proteins in the cell) prevents saturation of that permease with Enzyme III^{glc}. Consequently, under these conditions strong inhibition of inducer uptake by the PTS-mediated mechanism cannot occur. The net result is that as the number of permease proteins increases above the number of Enzyme III^{glc} molecules, the maximal percent inhibition decreases. When the permease is no longer rate-limiting for sugar uptake, the percent inhibition approaches zero. This postulate appears to account for the desensitization phenomenon. It is in agreement with the results and conclusions of Dills *et al.* (1980) and of Saier *et al.* (1982b).

The proposed mechanism of desensitization suggests that the level of Enzyme III^{glc} as well as those of the target permeases (and other cell constituents that bind Enzyme III^{glc}) may determine sensitivity to regulation. In this regard, it is worth noting that most reports concur that the level of Enzyme III^{glc} is either invariant or nearly invariant, regardless of growth conditions (Scholte *et al.*, 1981; Nelson *et al.*, 1982; Mitchell *et al.*, 1982; Saier and Roseman, 1976a; Saier and Feucht, 1975).

iii. Fine Structure Genetic Analysis of Regulatory Mutations

Until recently, the target permease-specific regulatory mutations were poorly characterized. However, fine-structure mapping of mutations that specifically abolish PTS-mediated regulation of the lactose permease in *E. coli* showed that two such mutations mapped within and to the C-terminal side of the *lacY* gene which codes for the lactose permease (R. Tuttle, M. Pfahl, and M. H. Saier, Jr., unpublished results, discussed in Saier, 1982). The *lac*-specific regulatory mutations have been termed *lacY^R*.

The recent demonstration that the C-terminus of the lactose permease is nonessential for catalytic turnover of the carrier (Bocklage and Müller-Hill, 1983; Griesser *et al.*, 1983) leads to the possibility that the lactose permease consists of two domains, an N-terminal domain, which is largely concerned with catalytic function, and a C-terminal domain, which is largely concerned with regulation. Further experimentation will be required to test this speculation.

Four independently isolated mutations that specifically abolished PTS-mediated regulation of the maltose permease in *E. coli* were found to map within the *malK* gene (M. Schwartz and M. H. Saier, Jr., unpublished results). Evidence for co-dominance of the mutant allele with the wild-type allele was obtained. These mutations have been termed *malK^R*. Employing a procedure developed by Bavoil *et al.* (1980), one mutant *malK^R* gene product was preliminarily found to exhibit an altered electrophoretic mobility, possessing the same size as the wild-type protein but increased acidity (T. Osumi, P. Bavoil, H. Nikaido, and M. H. Saier, Jr., unpublished results). As the *malK* gene product is an essential constituent of the maltose permease (see Chapter 2, Section D), these results suggest that the target permeases are subject to allosteric regulation by a regulatory protein that is encoded by the *crrA* gene. Missense mutations that appropriately alter the allosteric binding site of the permease so that the regulatory protein is not recognized would be expected to abolish permease regulation without affecting its transport function. This explanation provides insight into the presumed consequences of the permease-specific regulatory mutations discussed above.

The results described in the preceding paragraphs are consistent with a mechanism in which the product of the *crrA* gene, termed RPr for *R*egulatory *Pr*otein, interacts with the allosteric binding site of the target permease, thereby converting the permease to a conformation with low activity (Saier, 1977; Saier and Feucht, 1975). Phosphorylation of RPr (at the expense of phosphoenolpyruvate in a process catalyzed by Enzyme I and HPr) is thought to give rise to RPr-P, which either is not effective in binding to the allosteric site or else cannot induce the conformational change that reduces permease activity. This model, which is illustrated in Fig. 4.7, appears to account for most of the data that bear on the PTS-mediated regulatory process.

iv. Direct Binding of Enzyme III^{glc} to the Lactose Permease

Recently, direct biochemical evidence for a regulatory involvement of the glucose Enzyme III (presumed but previously not proven to be RPr) has been obtained. Four experimental approaches involved studies of (a) the effect of Enzyme III^{glc} on lactose uptake in membrane vesicles, (b) the direct binding of Enzyme III^{glc} to the *lac* permease protein in membrane fragments, (c) cooperativity of lactose–Enzyme III^{glc} binding to the *lac* permease *in vivo*, and (d) demonstration of Enzyme III^{glc} inhibition of lactose permease activity in reconstituted proteoliposomes. Additionally, the identification of Enzyme III^{glc} as the product of the *crrA* gene has been accomplished. These results will be described in the following paragraphs.

Partially purified Enzyme III^{glc} was shocked into *E. coli* membrane vesicles that possessed high activity of the lactose permease. Intravesicular (but not extravesicular) Enzyme III^{glc} inhibited lactose uptake, and in the presence of Enzyme I and HPr, inhibition was abolished by intravesicular phosphoenolpyruvate (Dills *et al.*, 1982). These results provided the first biochemical evidence for the involvement of Enzyme III^{glc} as RPr in the proposed regulatory process.

Further evidence supporting the involvement of Enzyme III^{glc} as the allosteric regulatory protein which binds to a site on the permease came from direct protein–protein binding studies (Osumi and Saier, 1982a,b). In these studies, advantage was taken of an *E. coli* strain which contained a multicopy plasmid carrying the *lacY* gene and greatly overproduced the lactose permease (Teather *et al.*, 1980). Membrane fragments isolated from this strain contained from 5 to 15% of their total protein as the lactose permease. Direct binding of Enzyme III^{glc} to these membrane fragments was studied under various conditions and in the presence or absence of the other proteins of the PTS by rapidly and quantitatively pelleting the membranes in an Airfuge ultracentrifuge. Soluble enzymes not bound to the membrane remained in the supernatant fraction which was removed. After the membranes were sedimented, the purified Enzyme III^{glc} (and other enzymes of the phosphotransferase system which became membrane-associated) were released from the membranes by extraction with a buffered solution containing 1 M NaCl. The activities of the enzymes were then measured in the salt extract after removal of the membranes.

Employing these conditions, small amounts of Enzyme III^{glc} were found to bind to the membrane fragments. In some experiments, HPr, but not Enzyme I, also associated with the membranes, and HPr binding was absolutely dependent on the presence of Enzyme III^{glc} (Osumi and Saier, 1982a). This result is not surprising since Enzyme III^{glc} has been shown to bind HPr with high affinity (Jablonski *et al.*, 1983). No binding of the PTS proteins was observed when synthesis of the lactose permease was not induced. The formation of a membrane-bound complex between the lactose permease, Enzyme III^{glc} and HPr suggests that the *lac* permease recognition site on the Enzyme III^{glc} differs from the HPr recognition site on the Enzyme III^{glc}. It is also worth noting that a functional complex of the soluble PTS enzymes involved in methyl α-glucoside phosphorylation has been shown to be present in association with the membranes of *E. coli* vesicles (Saier *et al.*, 1982a).

The presence of a substrate of the lactose permease markedly enhanced the binding of Enzyme III^{glc} to the membranes, and this effect was specific (Fig. 4.8). All substrates of the *lac* permease tested [including lactose, thio-β-D-digalactoside (TDG), thio-methyl-β-D-galactoside (TMG) and melibiose] stimulated binding more than fourfold over background activity, but sugars which did not bind to the permease were without effect (Osumi and Saier, 1982b). As expected, high affinity substrates of the *lac* permease promoted Enzyme III^{glc} binding at low concentration (half-maximal stimulation occurred at 15 μM thiodigalactoside) while low affinity substrates promoted Enzyme III^{glc} binding only at high concentration (half-maximal enhancement occurred at 15 mM lactose). Binding showed a pH dependency with an optimum at pH 6.0, as had previously been demonstrated for thiodigalactoside binding to the permease (Kennedy *et al.*, 1974). Under optimal conditions, as much as 50% of the Enzyme III^{glc} added to the membrane suspension became membrane-bound, showing that nearly stoichiometric amounts of Enzyme III^{glc} bound to the permease and that nonspecific or artifactual adsorption of the protein to the membranes could not account for the results. Isogenic control strains which were *lacY*$^-$ did not exhibit galactoside-promoted binding of Enzyme III^{glc} to the membranes.

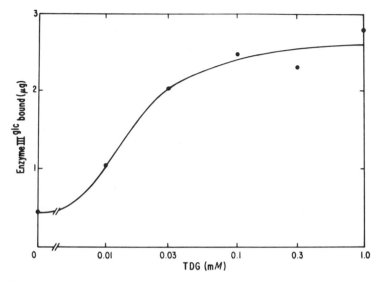

Fig. 4.8. Binding of Enzyme III^{glc} to membrane fragments isolated from *E. coli* strain T52RT which synthesizes elevated amounts of the lactose permease. Binding was plotted as a function of the thiodigalactoside (SGal$_2$) concentration. The binding experiments were carried out in the presence of thiodigalactoside at the concentrations indicated and according to the experimental procedures described in Osumi and Saier (1982a). [From Osumi and Saier (1982a), with permission.]

Earlier experiments had shown that the sulfhydryl reagent, *N*-ethylmaleimide, inactivated the lactose permease, and high-affinity substrates of the system such as thiodigalactoside protected against inactivation (Kennedy *et al.*, 1974). Accordingly, the effect of *N*-ethylmaleimide on Enzyme III^{glc} binding was studied in the presence and absence of thiodigalactoside (Osumi and Saier, 1982b). The reagent virtually abolished galactoside-promoted binding of Enzyme III^{glc}, but when thiodigalactoside was added prior to treatment with *N*-ethylmaleimide, the binding capacity was largely retained. These results substantiate the conclusion that binding of Enzyme III^{glc} to the membranes depends on an active configuration of the lactose permease.

The proposed model for the regulatory action of Enzyme III^{glc} on the lactose permease is illustrated in Fig. 4.7. When the intracellular pool of Enzyme III^{glc} is largely in the dephosphorylated form, binding of a substrate (lactose in the figure) to the permease induces the binding of Enzyme III^{glc}, and the ternary complex (lactose–lactose permease–Enzyme III^{glc}) exhibits low transport activity relative to the binary complex (lactose–lactose permease). When phosphorylated by the sequential reactions of the PTS, Enzyme III^{glc} dissociates from the lactose permease, and hence the permease assumes its active conformation.

The results of Osumi and Saier provided convincing evidence for the regulatory mechanism proposed in 1975 by Saier and Feucht. They demonstrate a direct interaction between Enzyme III^{glc}, the allosteric regulatory protein, and a target permease, in this case the *lac* permease. Although the concentrations of thiodigalactoside and of lactose which half-maximally promoted binding of Enzyme III^{glc} to the permease were significantly lower than the previously reported dissociation constants, this apparent anomaly can be explained by the cooperative effects of sugar and Enzyme III^{glc} binding. Thus, the affinity of the permease for the sugar substrates should be enhanced by the binding of Enzyme III^{glc} and vice versa.

The results described above have been confirmed and quantified (Nelson *et al.*, 1983). Moreover, the cooperative effect of Enzyme III^{glc} binding on galactoside binding, predicted from the cooperative effect of galactoside binding on Enzyme III^{glc} binding, was demonstrated. Approximately one molecule of Enzyme III^{glc} was reported to bind per molecule of lactose carrier. The K_D of the permease for Enzyme III^{glc} was estimated to be 10 ± 5 μM, and the binding of Enzyme III^{glc} was reported to result in a fourfold increase in the apparent affinity of the permease for the galactoside substrates.

Since phosphorylated Enzyme III^{glc} did not bind to the permease (Osumi and Saier, 1982b; Nelson *et al.*, 1983), the results support the conclusion, illustrated in Figure 4.7, that only free Enzyme III^{glc} binds to the permease to inhibit its activity. Consequently, the ternary complex, lactose–lactose permease–Enzyme III^{glc}, exhibits low transport activity relative to the binary complex of the sugar substrate with the permease (Osumi and Saier, 1982b). When phosphorylated by the sequential reactions of the PTS, Enzyme III^{glc} dissociates from the lactose permease, and the carrier assumes its active conformation (Fig. 4.8).

The estimated K_D of the lactose permease for Enzyme III^{glc} [10 μM as reported by Nelson *et al.* (1983)] is of considerable physiological interest. According to their purification data (Nelson *et al.*,

1983), Enzyme III^{glc} represents about 0.5% of the total soluble protein in the *E. coli* cell, a value which corresponds to an intracellular concentration of about 20–50 μM. In view of these values, it seems reasonable to assume physiological significance *in vivo* to the protein–protein interaction between Enzyme III^{glc} and the lactose permease measured in the *in vitro* assays. Moreover, maximal inhibition of lactose uptake by the PTS-mediated mechanism should not exceed 70–80%. The earlier uptake results showed that maximal inhibitory effects were of about this magnitude, and that residual activity could not be further depressed by addition of very high concentrations of sugar substrates of the PTS (Saier and Roseman, 1976b; Castro *et al.*, 1976; Saier *et al.*, 1978a). Since other uptake systems, such as those specific for glycerol and maltose, could be inhibited more than 90%, it must be assumed that these systems bind Enzyme III^{glc} with higher affinity (i.e., with a K_D of less than 10 μM).

v. Direct Regulation of Glycerol Kinase by Enzyme III^{glc}

Recently, evidence has been presented which suggests that the target of Enzyme III^{glc} action responsible for the inhibition of glycerol uptake is glycerol kinase, the first enzyme of glycerol metabolism, rather than the glycerol facilitator (Postma *et al.*, 1984). *In vitro* experiments were performed which showed that in crude extracts derived from glycerol grown wild-type cells of *Salmonella typhimurium* glycerol kinase was progressively inhibited by the addition of increasing concentrations of Enzyme III^{glc}. Half-maximal inhibition occurred at about 1 mg of Enzyme III^{glc} per ml (about 50 μM). This concentration of Enzyme III^{glc} was substantially in excess of that required for half-maximal inhibition of the lactose permease (about 200 μg/ml, or 10 μM) (Osumi and Saier, 1982a; Nelson *et al.*, 1983; Nelson and Postma, 1984). Since the intracellular concentration of Enzyme III^{glc} in *E. coli* and *S. typhimurium* is normally about 0.5–1.0 mg/ml (Scholte *et al.*, 1981), extrapolation of these *in vitro* results to the *in vivo* situation could lead to the prediction that glycerol uptake by whole cells could not be inhibited more than 50% by a sugar substrate of the PTS. Since inhibition in excess of 95% is frequently observed *in vivo,* it must be concluded that the *in vitro* assay conditions did not mimic the *in vivo* conditions. The enzyme in the crude extract may not have been in its native state, or a component of the extract may have inhibited binding of Enzyme III^{glc} to the kinase.

Recent studies with homogeneous glycerol kinase have provided some clarification regarding these uncertainties (M. J. Novotny *et al.,* 1985). Inhibition of glycerol phosphorylation by both of the two allosteric effectors of glycerol kinase, fructose 1,6-diphosphate and Enzyme III^{glc}, was found to be strongly pH dependent. For example, while fructose 1,6-diphosphate was inhibitory at neutral or slightly acidic pH values, it did not inhibit at basic pH values. Enzyme III^{glc} showed corresponding behavior, and both its affinity and its maximal inhibitory response were influenced by pH. At pH 7.0 in MES [2-(*N*-morpholino)ethanesulfonic acid] buffer, the K_i of the kinase for Enzyme III^{glc} was about 10 μM, while this value decreased to 4 μM when the pH was brought to 6.0. These values reflect much higher affinities than that measured by Postma *et al.* at pH 7.5 with a crude system. They also reflect higher-affinity binding than reported previously for the interaction between Enzyme III^{glc} and the lactose permease (Osumi and Saier, 1982b; Nelson *et al.,* 1983). The *in vitro* results obtained by Novotny *et al.* at pH 6 therefore account for the fact that the inhibitory effect of a PTS sugar on glycerol uptake is stronger than that on lactose uptake (Saier and Roseman, 1976b; Castro *et al.,* 1976). The *in vitro* conditions at pH 6 presumably mimic the *in vivo* conditions.

If glycerol kinase rather than the glycerol permease is the target of PTS-mediated inhibition of glycerol uptake, three predictions can be made. First, phosphorylation of Enzyme III^{glc} with phosphoenolpyruvate, Enzyme I and HPr should completely reverse the inhibitory effect of Enzyme III^{glc}; second, the activity of the facilitator should be insensitive to inhibition by Enzyme III^{glc}, and third, mutants which specifically render uptake of glycerol resistant to PTS-mediated regulation (Saier and Roseman, 1976a; Saier *et al.,* 1978a) should possess an altered glycerol kinase which is altered in the allosteric regulatory site of the enzyme which functions to bind the Enzyme III^{glc}. All three of these predictions have been verified (Postma *et al.,* 1984; M. J. Novotny, *et al.,* 1985). Since the mutations which render glycerol uptake insensitive to PTS-mediated regulation must map within the structural gene for glycerol kinase (*glpK*), these mutants should be designated *glpK*[R] by analogy with the *lacY*[R] and *malK*[R] mutants (see Section C,iii). It has been shown that *glpK*[R] mutants of *S. typhimurium* retain sensitivity to feedback inhibition by fructose 1,6-diphosphate (Saier *et al.*, 1978a). Further, *in vivo* assays indicated that *E. coli glpK*[i] mutants, which are insensitive to feedback inhibition by fructose 1,6-diphosphate, were still sensitive to PTS-mediated inhibition of glycerol uptake (Saier *et al.*, 1978a). It was therefore probable that the allosteric binding sites for Enzyme III[glc] and fructose 1,6-diphosphate on the surface of glycerol kinase are distinct.

This conclusion has been verified by *in vitro* assays (M. J. Novotny *et al.*, 1985). They isolated *glpK*[R] and *glpK*[i] mutants of *S. typhimurium* and assayed glycerol kinase for sensitivity to inhibition by both fructose 1,6-diphosphate and Enzyme III[glc]. Glycerol kinase isolated from a *glpK*[R] mutant was completely insensitive to regulation by Enzyme III[glc] at pH 6.5, the optimal pH for measuring regulatory interactions. However, the enzyme was fully sensitive to inhibition by fructose 1,6-diphosphate. By contrast, glycerol kinase in the *glpK*[i] mutant was insensitive to regulation by fructose 1,6-diphosphate but retained full sensitivity to regulation by Enzyme III[glc]. These studies established that the allosteric binding sites on glycerol kinase for fructose 1,6-diphosphate and Enzyme III[glc] are distinct. Some evidence suggests that these two sites may interact functionally (M. J. Novotny *et al.*, 1985). The recent cloning of the *glpK* gene (Conrad *et al.*, 1984) should facilitate genetic analyses leading to a detailed understanding of the structural and functional basis for PTS-mediated inhibition of glycerol uptake.

PTS-mediated regulation of glycerol kinase, rather than of the permease, is fully consistent with the inducer exclusion mechanism proposed previously (Saier and Feucht, 1975; Saier, 1977). This conclusion results from the fact that the inducer of the glycerol regulon is α-glycerophosphate. Inhibition of either entry or phosphorylation of glycerol should prevent accumulation of the inducer in the cytoplasm.

Because of the ease with which glycerol kinase can be prepared in crystalline form, it is likely that the detailed mechanism of its regulation by Enzyme III[glc] and the three dimensional structure of the glycerol kinase–Enzyme III[glc] complex will be elucidated more easily than the analogous processes and structures responsible for the regulation of the lactose, maltose, or melibiose permeases or of adenylate cyclase. For this reason it is both pertinent and interesting to review the known properties of this enzyme.

Escherichia coli glycerol kinase has been purified to homogenity, and many of its physicochemical and catalytic properties have been defined (Thorner and Paulus, 1971, 1973; Thorner, 1975). This tetrameric enzyme has a molecular weight of 220,000 with each identical subunit having a molecular weight of 55,000. It contains an excess of acidic amino acids, and glutamate is the only carboxyl terminal amino acid. All cysteine residues are present in a reduced form, and the enzyme is very sensitive to air oxidation and sulfhydryl reagents. Surprisingly, only 12 of the 20 sulfhydryl groups of the native enzyme are accessible for reaction with 5,5′-dithiobis(nitro-2-benzoic acid), and this number decreases to 2 in the presence of glycerol. Evidently, the binding of glycerol induces a substantial conformational change in the enzyme. In agreement with this conclusion, glycerol enhances the thermal stability of the enzyme in solution. The positive entropy of glycerol binding (+45 cal/mole/degree) also argues in favor of a substrate-induced conformational change (Thorner, 1975). It is possible that this conformational change enhances the affinity of the kinase for Enzyme III^{glc}.

Crystalline glycerol kinase exhibits broad substrate specificity, phosphorylating dihydroxyacetone and both D- and L-glyceraldehyde with low affinity (Thorner, 1975). While dihydroxyacetone is efficiently phosphorylated at high substrate concentration, L-glyceraldehyde, which is phosphorylated on the 3-position, also induces ATP hydrolysis to ADP and inorganic phosphate. Possibly the hydrated form of this triose is phosphorylated at position 1 to yield an unstable intermediate that decomposes to glyceraldehyde and phosphate. Only Mg-ATP can serve as the phosphoryl donor.

At pH 7.0 and 25°C, the apparent K_m values of glycerol kinase for its substrates, glycerol, dihydroxyacetone, D-glyceraldehyde, and L-glyceraldehyde are 10 μM, 0.5 mM, and 3 mM, respectively. Increasing the pH or decreasing the temperature increases the apparent K_m (decreases the affinity of the enzyme) for glycerol.

Even at saturating glycerol and Mg^{2+} concentrations, ATP saturation curves are not hyperbolic but yield double reciprocal plots with limiting slopes that indicate two apparent K_m values for Mg-ATP, one at 0.1 mM, the other at 0.5 mM. At an alkaline pH or in the presence of KCl, this apparently negative cooperativity is less pronounced, but under all conditions of assay, the effect is observed. It should be noted that the pH optimum of the enzyme is 9.5, with half-maximal activity at pH 7.0, and that KCl is slightly stimulatory.

At neutral pH, glycerol kinase from *E. coli* is subject to allosteric inhibition by fructose 1,6-diphosphate (Thorner and Paulus, 1973) and also to product inhibition by α-glycerol phosphate and ADP. K_i values for these three compounds are 0.5 mM, 0.5 mM, and 2 mM, respectively. While product inhibition is competitive with respect to the structurally related substrates, fructose 1,6-diphosphate inhibition is noncompetitive with respect to both substrates. Inhibition as a function of the allosteric effector is slightly sigmoidal, with greater than 80% of the activity being inhibited under optimal conditions. Inhibition is diminished at basic pH and high ionic strength. Each subunit appears to possess a single binding site for each of the two substrates as well as for the allosteric effector.

Work by de Riel and Paulus (1978a,b,c) has shown that the tetrameric glycerol kinase can dissociate into the dimeric species which is unable to bind fructose 1,6-diphosphate. Both the tetramer and the dimer have comparable catalytic activities, but only the tetramer binds the allosteric effector in a fashion which induces the enzyme to undergo a conformational transition to an inactive form (de Riel and Paulus, 1978a,b,c). Sensitivity of the tetrameric enzyme to allosteric inhibition by fructose 1,6-diphosphate may correlate with the apparently negative cooperativity with respect to ATP binding (see above). The relevance of these interesting properties to PTS-mediated regulation of the enzyme by the Enzyme III^{glc} has yet to be ascertained.

vi. Properties of Enzyme III^{glc}

Enzyme III^{glc} of *E. coli* and *S. typhimurium* has been independently purified to apparent homogeneity in three laboratories (Osumi and Saier, 1982b; Scholte *et al.*, 1981; Meadow and Roseman, 1982). The properties of the protein from these two organisms as reported by the three groups were essentially the same. It is a soluble, heat stable protein of molecular weight equal to about 20,000. Only 50% of its sugar phosphorylating activity was lost during incubation at 100°C for 1 hour (Meadow and Roseman, 1982). The protein exhibited partial hydrophobic characteristics and could be purified by hydrophobic chromatography on octylsepharose (Scholte *et al.*, 1981). Hydrophobic interactions may partially account for its high affinity for the integral membrane Enzyme II^{glc} (apparent $K_m \cong 3 \ \mu M$) as well as for the non-PTS permeases. The enzyme has also been shown to bind directly to HPr (Jablonski *et al.*, 1983). The nature of the forces responsible for these protein–protein interactions has not been investigated.

Enzyme III^{glc} was shown to lack the amino acids cysteine, tyrosine, and tryptophan, but it contained three histidyl residues (Meadow and Roseman, 1982). One of these histidyl residues appeared to become phosphorylated in the presence of phosphophenolpyruvate, Enzyme I, and HPr. This conclusion was based on the lability of the phosphoryl group to acid, its stability in solutions more basic than pH 10, and its sensitivity to hydrolysis at pH 8.5 and 37°C in the presence of 50 mM pyridine or 140 mM hydroxylamine. While both reagents hydrolyze phospho-histidyl residues in proteins as well as in phospho-Enzyme III^{glc}, only the latter reagent hydrolyzes acylphosphates. The rate of hydrolysis of phospho-Enzyme III^{glc} in acid was similar to that of phospho-Enzyme I in which the phosphoryl moiety is known to be linked at the N-3 position of the imidazole ring of a histidyl residue. This rate is much slower than that of phospho-HPr, in which the phosphoryl moiety is known to be linked at the N-1 position of the imidazole ring. N-3[^{32}P]phosphohistidine (9% of the total radioactivity in the protein) was identified after partial alkaline hydrolysis of phospho-Enzyme III^{glc} (Meadow and Roseman, 1982).

Although a single predominant species of Enzyme III^{glc} was present in actively growing bacteria, a slightly smaller protein, present in stationary phase cultures, co-purified with native Enzyme III^{glc}. The smaller form was shown to result from proteolytic cleavage of a heptapeptide from the N-terminus of the protein. While the cleavage product could accept the phosphoryl group from phospho-HPr, it transferred its phosphoryl moiety to methyl α-glucoside in the presence of partially purified Enzyme II^{glc} only 2–3% as rapidly as did native Enzyme III^{glc}. Because the cleavage product was largely absent from actively growing cells and because it did not appreciably inhibit the activity of native Enzyme III^{glc}, its physiological significance is questionable.

Compelling evidence now supports the conclusion that the *crrA* (*crr*) gene is the structural gene which encodes Enzyme III^{glc}

(Scholte *et al.*, 1982; Meadow *et al.*, 1982a,b; Cordaro *et al.*, 1976; Saier and Roseman, 1976a; M. J. Novotny and M. H. Saier, Jr., unpublished results). The evidence is as follows:

1. All *crrA* mutants of *S. typhimurium* exhibit low activity of the soluble Enzyme III^{glc} (Saier and Roseman, 1972, 1976a; Postma, 1982; Scholte *et al.*, 1982), and mutants deleted for the *crrA* gene lack this activity altogether (Cordaro *et al.*, 1976).

2. A mutant form of Enzyme III^{glc}, isolated from a *crrA* mutant, was purified and shown to exhibit abnormal aggregation properties, suggesting a structural change in the protein (Scholte *et al.*, 1982).

3. Other immunologically cross-reactive proteins, isolated from other *crrA* mutants, showed altered behavior during gel chromatography and native polyacrylamide gel permeation electrophoresis. These same proteins exhibited altered kinetics in the methyl α-glucoside phosphorylation assay (Meadow *et al.*, 1982a,b).

4. Selection of *crr* mutants of *E. coli* by integration of a defective bacteriophage Mu genome carrying most of the *lacZ* gene (encoding *E. coli* β-galactosidease), as well as a gene encoding resistance to ampicillin (Casadaban and Cohen, 1979), gave apparent *crrA–lacZ* gene fusions. The fused genes could be transcribed and translated *in vivo* to give fusion proteins which exhibited β-galactosidase activity, but little or no Enzyme III^{glc} activity. The fusion mutants obtained exhibited all of the properties of *crrA* mutants, and β-galactosidase activity became associated with soluble, constitutively synthesized proteins with molecular weights somewhat greater than that of native β-galactosidase (M. J. Novotny and M. H. Saier, Jr., unpublished results).

5. The *E. coli crrA* gene has been cloned into the high-copy-number plasmid, *p*BR322 (Meadow *et al.*, 1982b). This plasmid, designated pD545, was shown to carry a 1.3-kb insert. It was presumed to encode a functional *crrA* gene since it corrected the *crrA* mutant phenotype. More importantly, this plasmid directed the synthesis of a protein in *E. coli* maxicells which exhibited the same molecular weight as Enzyme III^{glc} and which cross reacted with anti-Enzyme III^{glc} antibody.

These five independent lines of evidence, particularly the results with the cloned gene, strongly argue that the *crrA* gene codes for (and does not merely influence the synthesis of) Enzyme III^{glc}.

A very interesting discovery of Scholte *et al.* (1982) was that *crrA* mutants which are devoid of the soluble Enzyme III^{glc} apparently contain membrane-bound Enzyme III^{glc} activity. Early results (Saier and Roseman, 1976a; Saier and Feucht, 1975) showed that *crrA* mutants of *S. typhimurium* with low activities of the soluble Enzyme III^{glc} (10–40% of wild type) still accumulated methyl α-glucoside at normal rates. Such an observation could be explained either if Enzyme III^{glc} is not rate limiting or if another protein can substitute for it. The membrane-bound protein with Enzyme III^{glc} activity reacted with and was inhibited by anti–Enzyme III^{glc} antibodies (Scholte *et al.*, 1982). Moreover, a mutant deleted for the *crrA* gene exhibited the same membrane-bound Enzyme III^{glc} activity, and this immuno-precipitable protein exhibited slightly higher mobility by SDS gel electrophoresis than did the soluble protein (Scholte *et al.*, 1982).

The presence of membrane-bound Enzyme III^{glc} activity has been confirmed (T. Osumi, unpublished results), and this activity was shown to remain with the membrane during extraction with urea and butanol by the procedure of Kundig and Roseman (1971; Saier *et al.*, 1977). These conditions quantitatively remove peripheral membrane proteins from the membrane fraction. It therefore appears that an integral membrane Enzyme III^{glc}, encoded by a gene which is distinct from the *crrA* gene, is present in *E. coli* and *S. typhimurium* cells. It catalyzes methyl α-glucoside phosphorylation, but whether or not it plays an important role in PTS-mediated regulation remains to be determined.

vii. *In Vivo* Cooperativity in Lactose/Enzyme III^{glc} Binding to the Lactose Permease

If cooperative binding of β-galactosides and Enzyme III^{glc} to the *lac* permease is an inherent characteristic of the regulatory process and not an artifact of the *in vitro* assay procedure, this cooperativity should be demonstrable *in vivo*. Assuming that the same regulatory protein, RPr (Enzyme III^{glc}), is responsible for the control of several uptake systems, those specific for glycerol, maltose, lactose, and melibiose, (Fig. 2.1), the binding of Enzyme III^{glc} to the lactose permease should render it unavailable for interaction with another system (i.e., glycerol kinase or the maltose permease). Further, if addition of thiodigalactoside or another substrate of the *lac* permease to the cell enhances the affinity of the *lac* permease for En-

zyme III^{glc}, Enzyme III^{glc} should be drained off of the other permeases, thereby relieving inhibition of their activities.

Experiments were therefore designed to test this hypothesis. The
E. coli strain, *T52RT,* which was used for the *in vitro* binding studies
(Osumi and Saier, 1982a,b), was used for the initial *in vivo* studies
(Saier *et al.,* 1983). Cells were grown in minimal-lactate plus glyc-
erol medium with a saturating concentration of the *lac* inducer,
isopropyl-β-thiogalactoside in order to induce high-level synthesis
of the lactose permease. Glucose was added 2 hours before harvest-
ing in order to induce synthesis of the glucose phosphotransferase.
Washed cells were then assayed for glycerol uptake. As shown in
Fig. 4.9, uptake of [^{14}C]glycerol was linear with time, but addition of

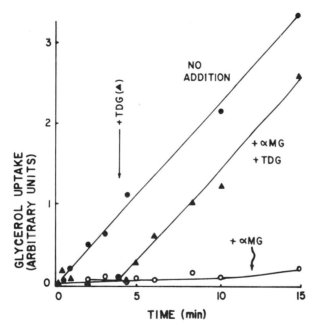

Fig. 4.9. Time course for the relief of PTS-mediated inhibition of glycerol uptake by
thiodigalactoside (TDG) in *E. coli* strain T52RT, which overproduces the lactose permease. The figure shows glycerol uptake in the absence of PTS-mediated inhibition (●);
the inhibitory effect of methyl α-glucoside (αMG) (○); and the relief of inhibition caused
by the addition of a saturating concentration of thiodigalactoside (TDG) (▲). The concentrations of methyl α-glucoside and thiodigalactoside were 1 and 0.1 mM, respectively.
[From M. H. Saier *et al.* (1983), with permission.]

the nonmetabolizable glucose analogue, methyl α-glucoside, resulted in immediate and virtually complete inhibition. This inhibition is attributed to dephosphorylation of phospho-Enzyme III^{glc} upon addition of the glucoside. If saturating amounts of thiodigalactoside (TDG) were subsequently added to the cell suspension, the inhibitory effect of the glucoside on glycerol uptake was immediately and quantitatively relieved (Fig. 4.9). Relief from inhibition presumably resulted from competition of the two uptake systems for Enzyme III^{glc}. Similar observations were made with the maltose permease (Saier *et al.*, 1983).

The effect observed with *E. coli* strain T52RT was not observed in a strain which lacked or contained normal levels of the lactose permease, but growth of wild-type *E. coli* in the presence of isopropyl-β-thiogalactoside plus cyclic AMP resulted in enhanced synthesis of the lactose permease so that galactosides relieved inhibition of glycerol uptake. Thiodigalactoside also relieved the inhibition of glycerol uptake caused by the presence of other PTS substrates such as fructose, mannitol, glucose, 2-deoxyglucose, and 5-thioglucose. Cooperative binding of sugar and Enzyme III^{glc} to the melibiose permease in *Salmonella typhimurium* was demonstrated, but no cooperativity was noted with glycerol and maltose. These results are consistent with a mechanism of PTS-mediated regulation of the lactose and melibiose permeases involving a fixed number of allosteric regulatory proteins (Enzyme III^{glc}), which may be titrated by the increased number of substrate-activated permease proteins. This work suggests that the cooperativity in the binding of sugar substrate and Enzyme III^{glc} to the permease, demonstrated previously in *in vitro* experiments, has mechanistic significance *in vivo*. It substantiates the conclusion that PTS-mediated regulation of non-PTS permease activities involves direct allosteric interaction between the permeases and Enzyme III^{glc}, the presumed regulatory protein of the PTS.

viii. Reconstitution of Enzyme III^{glc}–Mediated Regulation of the Lactose Permease in Proteoliposomes

In subsequent experiments the purified lactose permease was reconstituted in an artificial phospholipid bilayer according to the procedure of Newman *et al.* (Newman and Wilson, 1980; Newman *et al.*, 1981). Inhibition of [^{14}C]lactose counterflow was demonstrated upon addition of purified Enzyme III^{glc} to the proteoliposomes (Table 4.1; M. J. Newman *et al.*, unpublished results). Increasing concentrations of Enzyme III^{glc} resulted in increasing degrees of inhibition with the maximal inhibitory response equal to about 60% (Table 4.1). The permease might be expected to insert randomly into the proteoliposomes, with about 50% of the proteins in one orientation and the other 50% in the opposite orientation. Since all permease proteins would be expected to exhibit comparable counterflow activity regardless of orientation, but only those with their cytoplasmic side facing the extravesicular medium would be exposed to the exogenously added Enzyme III^{glc}, one would expect about 50% maximal inhibition. The slightly greater inhibitory response may be attributable to an unequal distribution of the protein between its two possible orientations during insertion into the liposomes. Such an unequal distribution has been demonstrated in some liposome preparations (H. R. Kaback, personal communication).

TABLE 4.1

Inhibition of [^{14}C]Lactose Counterflow by Purified
Enzyme III^{glc} in Reconstituted Proteoliposomes
Containing Homogeneous Lactose Permease[a]

lac Permease proteoliposomes	III^{glc} (mg/ml)	Counterflow (%)
+	0	100
+	0.5	70
+	1.0	43
+	1.5	43
+	2.0	40
−	0	10

[a] Proteoliposomes containing the homogeneous lactose permease protein were prepared, and [^{14}C]lactose counterflow was assayed as described by Newman and Wilson (1980) and Newman *et al.* (1981). Increasing amounts of Enzyme III^{glc}, purified according to the procedure of Meadow and Roseman (1982), were added to the extravesicular fluid before counterflow activity was assayed. No inhibitory response was observed when a mutant Enzyme III^{glc} protein isolated from a *crrA* mutant of *S. typnimurium* was added to the liposomes. (M.J. Newman, T. Osumi, H.R. Kaback, N.D. Meadow, S. Roseman, and M.H. Saier, Jr., unpublished results.)

D. REGULATION OF PTS-MEDIATED CARBOHYDRATE
UPTAKE BY COMPETITION FOR PHOSPHO-HPr

In a previous article (Dills *et al.*, 1980), evidence was presented supporting the hypothesis that different Enzyme II-Enzyme III complexes of the PTS compete for the common phosphoryl donor, phospho-HPr. Such a regulatory mechanism is illustrated in Fig. 4.10. As a consequence of this proposed mechanism, the uptake of one PTS sugar can inhibit uptake of another. The characteristics of this type of inhibition are as follows:

1. Uptake of one PTS sugar is maximally inhibited immediately upon addition of a second PTS sugar. There is no lag period for inhibition, and the degree of inhibition does not correlate with the intracellular concentration of metabolites derived from the inhibiting sugar.

2. The intensity of inhibition is increased (a) if the cells are energy depleted so that the internal concentration of phosphoenolpyruvate is diminished, (b) if the level of active Enzyme I is reduced (i.e., by mutation) so that the rate of HPr phosphorylation decreases, or (c) if the level of active HPr is reduced so that the maximal concentration of phospho-HPr is diminished.

3. Any sugar substrate of the PTS can inhibit the uptake of any other PTS sugar, provided that the Enzyme II (or Enzyme II–Enzyme III pair) responsible for the uptake of the inhibiting sugar is present in sufficient amounts. Representative data documenting this fact (Table 4.2) show that the inhibiting PTS sugar inhibits most strongly when the Enzyme II for that sugar is induced while that for the inhibited sugar is not.

4. *crrA* mutations in the genetic background of *S. typhimurium* strain *ptsI17,* which abolish regulation of non-PTS permeases by sugar substrates of the PTS (Saier and Roseman, 1976a), had essentially no effect on the inhibition of fructose uptake by methyl α-glucoside.

5. In an *E. coli* genetic background (Amaral and Kornberg, 1975; Kornberg and Watts, 1978), a *"crr"* mutation decreased the rate and extent of methyl α-glucoside uptake, enhanced the degree of inhibition of methyl α-glucoside uptake by fructose, and decreased the inhibition of fructose uptake by methyl α-glucoside. This effect

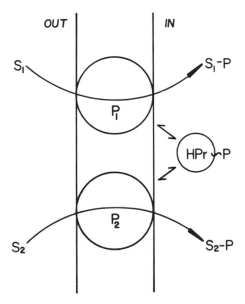

Fig. 4.10. Inhibition of sugar uptake via the phosphotransferase system by competition for phospho-HPr. The figure shows two permeases, P_1 and P_2, which are Enzymes II of the phosphotransferase system. These proteins transport and phosphorylate their exogenous sugar substrates, S_1 and S_2, respectively. When both S_1 and S_2 are present in the extracellular medium, the relative rates of S_1 and S_2 transport are determined both by the affinities of the two permeases for phospho-HPr and by their relative quantities (see Table 4.2).

was specific to glucose and methyl α-glucoside and was attributed to decreased affinity of the Enzyme II^{glc}–Enzyme III^{glc} complex for phospho-HPr (M. H. Saier, Jr., P. Watts, and H. L. Kornberg, unpublished results).

The conclusion that phospho-HPr competition is a physiologically significant mechanism regulating sugar uptake via the PTS in *S. typhimurium* has been independently substantiated (Scholte and Postma, 1981b). In these studies, thio β-D-glucoside and 5-thio-D-glucose were used as substrates of the phospho-HPr:glucose phosphotransferase while mannose was used as the sugar substrate of the phospho-HPr:mannose phosphotransferase. Under the conditions

TABLE 4.2

Effect of Enzyme II Induction on the Inhibition of PTS Sugar Uptake by S.
typhimurium Strain ptsI17[a]

[^{14}C]Sugar substrate	Carbon source for growth	Inhibitory sugar	Uninhibited uptake rate (μmoles/min/ g dry wt.)	Inhibition (%)
Mannitol	Mannitol	Fructose	2.3	0
Mannitol	Fructose	Fructose	1.2	42
N-acetylglucosamine	N-acetylglucosamine	Mannitol	2.8	20
N-acetylglucosamine	Mannitol	Mannitol	2.0	75

[a] The growth medium consisted of half-strength Medium 63 containing 16 g of nutrient broth and the inducing sugar (indicated above) at a concentration of 0.5%. S. typhimurium, strain ptsI17 cells, grown at 37°C, were harvested during logarithmic growth, washed three times, and resuspended to a cell density of 0.12 mg, dry weight per milliliter in half-strength Medium 63. Uptake of ^{14}C-substrates was measured at 30°C as described previously (Saier et al., 1976b). The concentration of the ^{14}C-sugar substrate was 10 μM while that of the inhibitory sugar was 1 mM.

used, the activity of the mannose phosphotransferase was shown to be strongly inhibited by the nonmetabolizable substrates of the glucose phosphotransferase. In agreement with the results discussed above, it was concluded that the flow of phosphoryl groups through Enzyme I and HPr can be rate-limiting for sugar uptake in intact S. typhimurium cells. It is interesting to note that while both the mannose and glucose phosphotransferase activities are inducibly synthesized in certain strains of both E. coli (Kornberg and Reeves, 1972; Saier et al., 1976a), and S. typhimurium (Rephaeli and Saier, 1980b), these enzyme activities have been reported to be constitutively synthesized in other strains of these species (Scholte and Postma, 1981b; Kornberg and Reeves, 1972).

Recently, Dills and Seno (1983) have examined the regulation of hexitol catabolism by glucose and 2-deoxyglucose in the gram-positive bacterium, Streptococcus mutans. Their analyses showed that the hexoses not only inhibited uptake of the hexitols by a mechanism which exhibited all of the characteristics of phospho-HPr competition, but that the resultant exclusion of the hexitols from the cells resulted in strong repression of the synthesis of the hexitol catabolic enzymes (Dills and Seno, 1983). This result suggests that competition for phospho-HPr, as illustrated in Fig. 4.10, can play an important role in the regulation of carbohydrate catabolic enzyme synthesis. However, the possible involvement of a protein kinase-mediated regulatory mechanism cannot be ruled out at this time (see Chapter 6, Section C).

5

Mechanisms of Adenylate Cyclase Regulation in Gram-Negative Bacteria

Cyclic AMP was first identified in *E. coli* by Makman and Sutherland in 1965. This cyclic nucleotide is synthesized by a cytoplasmic adenylate cyclase, which is frequently found in association with the envelope fraction of disrupted cells. Cyclic AMP is degraded to 5'-adenosine monophosphate by a cytoplasmic cyclic nucleotide phosphodiesterase and transported out of the cell by a chemiosmotically driven permease system (Fig. 5.1). Although it is widely accepted that this cyclic nucleotide is a principal mediator of the phenomenon of catabolite repression (see Chapter 4), repressive phenomena have been demonstrated in the absence of both cyclic AMP and its recep-

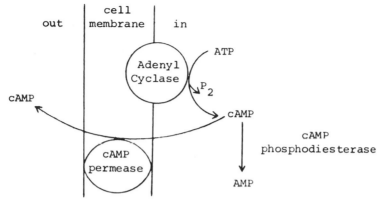

Fig. 5.1. Metabolism of cyclic AMP in enteric bacteria. Cyclic AMP is synthesized by adenylate cyclase, which is a soluble protein that may be associated with the cytoplasmic membrane in the intact cell. Intracellular cyclic AMP can be degraded to 5'-AMP by cyclic AMP phosphodiesterase, or it can be transported out of the cell by an energy-dependent cyclic nucleotide efflux system. [From Saier (1979).]

tor proteins (Guidi-Rontoni *et al.*, 1980, 1982). Mutants lacking the cyclic AMP phosphodiesterase have been isolated and characterized in both *S. typhimurium* and *E. coli* (Alper and Ames, 1975; Monard *et al.*, 1969), and some of the properties of the enzyme have been described (Brana and Chytel, 1966; Rickenberg, 1974). Mutants which lack this enzyme show elevated levels of cyclic AMP and are partially resistant to repression of enzyme synthesis (Rickenberg, 1974; Saier and Feucht, 1975; Feucht and Saier, 1980). There is no convincing evidence that the regulation of this enzyme mediates the physiologically relevant fluctuations in cellular cyclic AMP concentrations in response to changes in growth conditions. The low affinity of the enzyme for cyclic AMP suggests that its primary function is to protect against the toxic effects of high cytoplasmic concentrations of the cyclic nucleotide (Judewicz *et al.*, 1973; M. H. Saier, Jr., unpublished observations).

The efflux of cyclic AMP is mediated by a cyclic nucleotide-specific transport system, which is energized by the membrane potential (Saier *et al.*, 1975b; Fraser and Yamazaki, 1979). The efflux process was studied by Goldenbaum and Hall (1979) using everted membrane vesicles. The affinity of the system for cyclic AMP was low ($K_m = 10$ mM), and efflux was inhibited by a number of cyclic AMP analogues. Vesicles from a cyclic AMP receptor protein-negative (*crp*) mutant showed about threefold elevated cyclic AMP transport activity relative to vesicles from the parental strain. This observation correlates with reports of others that *crp* mutants show increased rates of cyclic AMP efflux (Potter *et al.*, 1974; Fraser and Yamazaki, 1978). Conversely, a *crp** mutation, which *promotes* transcription of catabolite-sensitive operons in the absence of cyclic AMP, was shown to inhibit the export of cyclic AMP without altering adenylate cyclase activity. This mutant strain showed a threefold increase in the concentration of cytoplasmic cyclic AMP, while exhibiting a ninefold decrease in extracellular cyclic AMP (Nelson *et al.*, 1982). These results clearly implicate the cyclic AMP receptor protein in regulation of the activity or synthesis of the cyclic AMP transport system.

Regardless of whether it regulates synthesis or the activity of the cyclic AMP permease, the cyclic AMP–cyclic AMP receptor protein complex presumably functions as a negative effector. Consequently, *crp⁻* mutants, which lack the cyclic AMP receptor protein, exhibit elevated activity of the cyclic AMP transporter, while *crp** mutants, which have the cyclic AMP receptor protein locked in the active configuration, possess depressed activity. In view of the known function of the cyclic AMP receptor protein in transcriptional initiation, a mechanism involving regulation of the synthesis rather than the activity of the permease seems likely.

As for the cyclic nucleotide permease, adenylate cyclase synthesis or activity appears to be regulated in a negative sense by the cyclic AMP-cyclic AMP receptor protein complex. This suggestion is based on the fact that *crp⁻* mutants exhibit elevated adenylate cyclase activity as revealed by several studies (Rephaeli and Saier, 1976; Fraser and Yamazaki, 1979; Majerfeld *et al.*, 1981). Recently, the *cya* locus encoding adenylate cyclase has been cloned, and the associated regulatory region was sequenced (Roy and Danchin, 1982; Roy *et al.*, 1983a,b). The *cya* gene was found to be preceded by two promoters of nearly equivalent strength. These two promoters, designated P1 and P2, were both shown to be active *in vivo*. Employing *cya–lacZ* gene fusions, P1 activity appeared to be sensitive to catabolite repression. However, transcription initiated at the stronger promoter, P2, did not respond to glucose. No effect of cyclic AMP or its receptor protein was demonstrable although the DNA sequence revealed a probable binding site for the protein downstream from the P2 promoter (Roy *et al.*, 1983b). It is not clear how these observations relate to the *in vivo* observations concerning the effects of *crp* mutations on net cyclic AMP synthesis (Rephaeli and Saier, 1976; Fraser and Yamazaki, 1979).

The activity of adenylate cyclase is subject to stringent regulation. The addition of glucose or another metabolizable sugar results in the immediate inhibition of its activity. Consequently, the cyclic AMP biosynthetic enzyme and the cyclic AMP efflux system appear to operate in conjunction to lower cellular cyclic AMP concentrations in response to energy availability. While adenylate cyclase is regulated by various energy sources, as discussed in this chapter, the efflux system is activated by the proton electrochemical gradient. High cyclic AMP levels, in response to energy starvation, allow enhanced rates of synthesis of carbohydrate catabolic enzyme systems, thereby preparing the cell for the metabolism of an exogenously supplied carbon source. By contrast, energy sufficiency lowers cyclic AMP levels so that further synthesis of carbohydrate catabolic enzyme systems will be inhibited. The net result is prevention of the synthesis of carbohydrate catabolic enzymes and transport systems in excess of the needs of the cell.

Recent evidence has shown that in enteric bacteria, such as *E. coli* and *S. typhimurium*, there are at least three distinct mechanisms by which the activity of adenylate cyclase is regulated. All of these mechanisms apply to the regulation of inducer uptake as well (Fig. 5.2). While these regulatory mechanisms are not understood at the molecular level, adenylate cyclase has been purified, and the monomeric soluble protein was shown to possess a molecular weight of 95,000 (Yang and Epstein, 1983). Genetic experiments with the cloned *cya* gene suggest that only the N-terminal half of the protein is required for catalytic activity (Roy and Danchin, 1982; Wang *et al.*, 1981). This result suggests that the protein may consist of two domains. The N-terminal domain is concerned with catalysis of cyclic AMP synthesis, while the C-terminal domain is involved in regulation of the catalytic domain. Indeed, the truncated protein is at least partially resistant to regulation by glucose (Roy *et al.*, 1983b; J. Y. J. Wang, personal communication; see Section C).

As proposed earlier (Saier, 1979), each of the energy sources which can inhibit permease activity and thereby control rates of inducer uptake, also regulate adenylate cyclase activity. Thus, adenylate cyclase is apparently responsive to fluctuations in the chemiosmotic energy level of the cell, the cytoplasmic concentra-

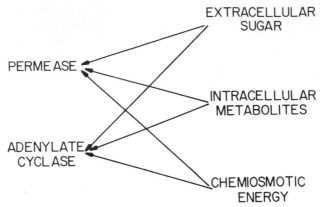

Fig. 5.2. Schematic depiction of regulatory agents that control the activities of adenylate cyclase and carbohydrate permeases in bacteria. Extracellular sugar substrates of the PTS control these activities by the PTS-mediated mechanism and also by simple competition at the extracellular sugar binding site of the permease. Intracellular metabolites and chemiosmotic energy function in regulatory capacities by quite distinct mechanisms. [From Saier (1982), with permission.]

tions of metabolites, and the extracellular concentrations of certain sugars. Coordinate regulation of the concentrations of the two cytoplasmic molecules (inducer and cyclic AMP), which control the rates of transcriptional initiation of carbohydrate catabolic operons, provides the cell with a dual mechanism controlling transcription. Bacteria have apparently evolved a safety valve to insure that target proteins will only be synthesized when required for the acquisition of cellular carbon and the generation of energy. In this chapter, the evidence for multiple mechanisms of adenylate cyclase regulation in enteric bacteria will be reviewed.

A. REGULATION OF ADENYLATE CYCLASE BY THE MEMBRANE POTENTIAL

Work of Peterkofsky and Gazdar (1979) has led to the proposal that the activity of adenylate cyclase is regulated by the proton electrochemical gradient (proton motive force) in *E. coli*. Collapse of the proton motive force (pmf), which results from coupled unidirectional lactose:H^+ influx via the *lac* carrier or dissipation of the pmf with a proton-conducting uncoupler such as carbonylcyanide-*m*-chlorophenylhydrazone (CCCP), results in inhibition of adenylate cyclase activity *in vivo*.

In toluenized *E. coli* cells which are leaky to small molecules, inhibiton of adenylate cyclase by lactose was dependent on β-galactosidase, presumably because the glucose released during lactose hydrolysis could elicit PTS-mediated control of the enzyme (see Section C). In contrast, with intact cells, nonmetabolizable lactose analogues such as isopropyl-β-thiogalactoside (IPTG) partially inhibited adenylate cyclase. These analogues were not inhibitory in toluenized cells or in intact cells containing a nonfunctional *lac* permease. Employing an energy-uncoupled mutant of *E. coli* in which lactose translocation was uncoupled from proton transport (Wilson and Kusch, 1972), methyl β-thiogalactoside (TMG), another nonmetabolizable galactoside which inhibited adenylate cyclase in the wild-type organism, was without effect. This result was interpreted by assuming that proton influx rather than sugar transport is responsible for the inhibition of adenylate cyclase.

If proton influx inhibits adenylate cyclase by neutralizing the proton motive force, proton-conducting uncouplers should exert a similar effect. In fact, 4 μM carbonylcyanide-*m*-chlorophenylhydrazone strongly inhibited adenylate cyclase under conditions which did not appreciably lower cellular concentrations of ATP. Comparison of the concentrations of CCCP which inhibited uptake of TMG and cyclic AMP synthesis revealed that the concentration of the inhibitor which depressed the activities of both systems by 50% was 1 μM, and the two curves were superimposable within experimental error. Thus, the rates of TMG uptake and cyclic AMP synthesis were apparently coordinately inhibited by dissipation of the proton electrochemical gradient.

The mechanism of adenylate cyclase regulation by the pmf is unknown. Peterkofsky and Gazdar proposed the involvement of a factor, distinct from adenylate cyclase, which senses the energy state of the cell. On the other hand, a mechanism analogous to that proposed by Robillard (1982) for the regulation of lactose permease function by reversible sulfhydryl oxidation can be contemplated. It is also possible that the enzyme merely senses the cytoplasmic pH. Further work will be required to determine the regulatory mechanism involved.

B. REGULATION BY INTRACELLULAR SUGAR PHOSPHATES

It has been shown that intracellular sugar phosphates strongly inhibit adenylate cyclase under conditions which result in no apparent change in the membrane potential or cellular ATP levels, and where extracellular sugar is essentially absent (J. D. Desai and M. H. Saier, Jr., unpublished results). Because these experiments provide evidence for an additional and possibly very general regulatory mechanism, they will be presented here.

2-Deoxyglucose is accumulated in the cell cytoplasm as a result of the group translocation process catalyzed by the phosphotransferase system (Rephaeli and Saier, 1980a). Uptake is time and concentration dependent, as shown in Fig. 5.3. The intracellular concentration of the radioactive sugar phosphate plateaued after about a 4-min incubation period and was relatively stable. As shown in Fig. 5.4, 2-deoxyglucose accumulation followed saturation kinetics with an apparent K_m of about 2 μM under the conditions employed (inset to Fig. 5.4). More than 85% of the accumulated [^{14}C]sugar was shown to be 2-deoxyglucose 6-phosphate rather than the free sugar. Increasing concentrations of cytoplasmic 2-deoxyglucose 6-phosphate strongly inhibited adenylate cyclase even though the extracellular concentration of 2-deoxyglucose was negligible under the assay conditions. Assays for cytoplasmic ATP (Chapman et al., 1971) and the initial rates of serine and proline uptake [indirect measures of the membrane potential (Saier et al., 1975b)] revealed that neither quantity was appreciably affected by the accumulated 2-deoxyglucose 6-phosphate (Fig. 5.4). Serine and proline uptakes were shown to be completely blocked by 25 μM carbonylcyanide-m-chlorophenylhydrazone, which effectively dissipates the membrane potential in these cells (Saier et al., 1975b). Thus, inhibition of adenylate cyclase cannot be attributed to cellular ATP depletion or to either of the two established regulatory mechanisms discussed in Sections A and C of this chapter. An effect of intracellular 2-deoxyglucose 6-phosphate on adenylate cyclase activity must therefore be proposed.

Recently it has been shown that in *Streptococcus pyogenes* glycolytic intermediates stimulate a protein kinase-catalyzed reaction which results in the phosphorylation of a seryl residue in HPr, a high-energy phosphoryl carrier protein of the bacterial phosphotransferase system (Chapter 6, Section C). Further, it has been shown that gram-negative bacteria possess several protein kinases which phosphorylate at least 10 cellular proteins (see Chapter 6,

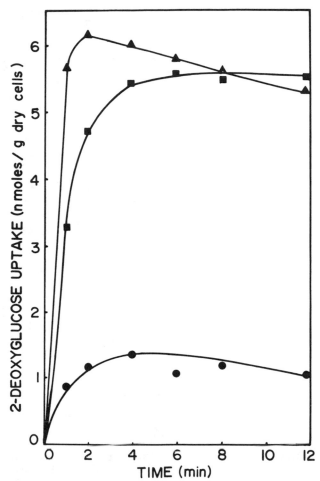

Fig. 5.3.   Time course for uptake of 2-deoxy[^{14}C]glucose by *S. typhimurium* strain LJ62 (*cpd-401*). Cells were prepared for transport studies and then preincubated at 37°C for 5 min with aeration before uptake was initiated by addition of 2-deoxy[^{14}C]glucose. At indicated time intervals, cells were removed for filtration, and the extent of 2-deoxyglucose uptake was determined. Concentrations of 2-deoxy[^{14}C]glucose were as follows: ●, 1μM; ■, 10 μM; ▲, 100 μM. The experiment was performed by Dr. J. D. Desai in the author's laboratory.

Sections D and E). These observations lead to the possibility that protein phosphorylation, via an ATP-dependent mechanism, is responsible for the inhibitory effect of cytoplasmic 2-deoxyglucose 6-phosphate on adenylate cyclase and several permease proteins as discussed in Section B of Chapter 4. Further experimentation will be required to determine the veracity of this working hypothesis.

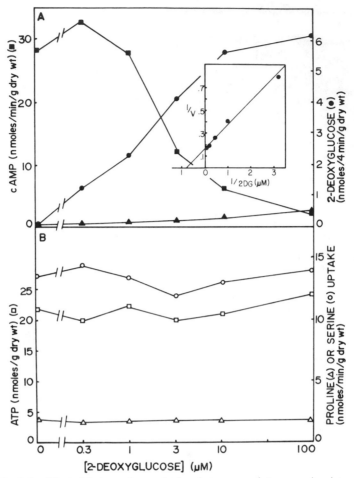

Fig. 5.4. Effect of 2-deoxyglucose 6-phosphate accumulation on adenylate cyclase activity, ATP levels, and serine and proline uptake rates in intact cells of *S. typhimurium* strain LJ62 (*cpd-401*). Cells were exposed to the indicated concentration of 2-deoxyglucose for 4 min at 37°C and were then washed three times with Medium 63 by centrifugation at 0°C. Cells were resuspended in Medium 63 to a final cell density of 0.4 mg dry cells per milliliter. The levels of cyclic AMP were determined at 0 time and after a 15-min incubation period at 34°C. (A, ■). ATP was estimated after a 15-min incubation period at 37°C (B, □). In both cases, cells were heated to 100°C for 2–3 min and removed by centrifugation before total cyclic AMP and ATP were determined in the supernatant fluids. The ability of cells to take up serine (B, ○) and proline (B, △) was used as a measure of the magnitude of the membrane potential (Lombardi and Kaback, 1972). The inset shows a Lineweaver-Burk plot of 2-deoxyglucose uptake. The calculated apparent K_m value was 2.2 μM. The intracellular levels of 2-deoxyglucose (A, ●) were determined after a 10-min incubation period at 37°C, following washing. The experiment was performed by Dr. J. D. Desai in the author's laboratory.

C. PTS-MEDIATED REGULATION OF ADENYLATE CYCLASE

The involvement of the phosphotransferase system in the regulation of adenylate cyclase and in the consequent transcription of carbohydrate catabolic enzyme systems was established in numerous studies (Saier and Feucht, 1975; Feucht and Saier, 1980; Glesyna *et al.*, 1983; Peterkofsky and Gazdar, 1974, 1975). Representative results are shown in Fig. 5.5 and 5.6. These figures show the amounts of cyclic AMP which are synthesized and secreted into the culture media of representative isogenic *E. coli* strains when these cells are grown in a mineral salts medium containing pyruvate and methyl α-glucoside. The bacterial strains studied included a wild-type strain (Fig. 5.5 A and B); a cyclic AMP phosphodiesterase-

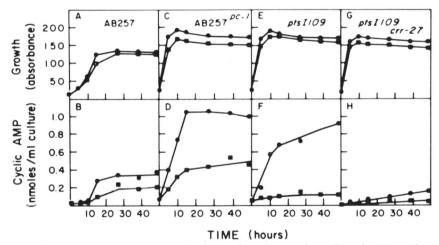

Fig. 5.5. Net production of cyclic AMP in the culture medium of *E. coli* strains with (■) and without (●) 0.1% methyl α-glucoside, as a function of time. Aliquots of cells grown in Medium 63 containing 0.5% sodium pyruvate and 20 μg of L-methionine per milliliter were transferred (without washing) to fresh medium of the same composition (20 ml in 125 ml Erlenmeyer flasks equipped with side arms for turbidimetric determination). Aliquots (2ml) were periodically removed for determination of total cyclic AMP. An isogenic series of mutant strains was used as follows: Parts A and B: SB2251 (AB257), *met* (the parental, wild-type strain); C and D: SB2252 (AB257PC-1), *met cpd-454* (a cyclic AMP phosphodiesterase-negative strain derived from strain AB257); E and F: SB2260, *met cpd-454 ptsI109* (a mutant strain which, in addition to lacking cyclic AMP phosphodiesterase, was deficient for but not totally lacking Enzyme I of the PTS); G and H: SB2876, *met cpd-454 ptsI109 crrA27* (an Enzyme III^{glc}–deficient strain isolated from strain SB2260).

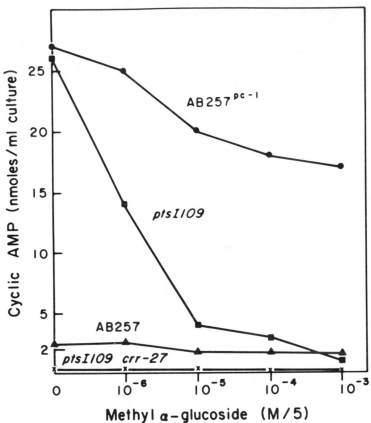

Fig. 5.6. Net synthesis of cyclic AMP by *E. coli* strains as a function of methyl α-glucoside concentration. The experiment was conducted as in Fig. 5.5 except that flasks contained 10 ml of medium with the indicated concentration of methyl α-glucoside. After swirling at 37°C for 40 hours, cells were in stationary phase, and cyclic AMP synthesis had ceased. Total cyclic AMP was determined as reported previously (Feucht and Saier, 1980). The bacterial strains used were the same as those described in Fig. 5.5.

negative (*cpd*) strain (Parts C and D); an Enzyme I-deficient (*ptsI*) strain derived from the *cpd* strain (Parts E and F); and a *cpd*, *ptsI*, *crrA* triple mutant (Parts G and H). The top figures show growth, while the bottom figures shown cyclic AMP production. Several conclusions can be drawn from the data presented in Figs. 5.5 and 5.6: (1) Net synthesis of cyclic AMP ceases shortly after cells reach the stationary growth phase. (2) The net production of cyclic AMP is substantially elevated by introduction of a phosphodiesterase-negative mutation (compare strain AB257 with strain AB257^{PC-1}), and the amount of cyclic AMP produced by the Enzyme I-deficient strain (*ptsI109*) in the absence of sugar was approximately the same. (3) Methyl α-glucoside inhibited cyclic AMP production in the cyclic AMP phosphodiesterase-negative mutant, and the *ptsI109* mutation enhanced sensitivity to this effect. (4) Inhibition of adenylate cyclase was not a transient effect, but lasted for the duration of the experiment. (5) The *crrA27* mutation decreased the quantity of cyclic AMP produced to a level below that observed for the wild-type strain.

Methyl α-glucoside was not the only sugar which affected net cyclic AMP production. In the parental phosphodiesterase-negative strain, AB257^{PC-1}, only those sugar substrates of the PTS which were effective inducers of the corresponding Enzyme II complexes (Saier *et al.,* 1976a) effectively depressed net cyclic AMP production (Table 5.1, column 1). By contrast, *all* sugar substrates of the PTS were potent inhibitors of adenylate cyclase in the leaky Enzyme I mutant (Table 5.1, column 2). Sugars which were not transported and phosphorylated by the PTS were less effective in depressing rates of cyclic AMP production under the conditions of the experiment.

More extensive genetic and physiological studies were carried out using a variety of "tight" and "leaky" mutants defective in the following genes: *ptsI* (Enzyme I⁻); *ptsH* (HPr⁻); *cpd* (cyclic AMP phosphodiesterase⁻); *crrA* (Enzyme III^{glc}⁻); *cya* (adenylate cyclase⁻); *crp* (cyclic AMP receptor protein⁻); and genes encoding selected Enzymes II of the PTS (Saier *et al.,* 1976b; Feucht and Saier, 1980).

The essential features of the mechanism by which the PTS regulates the activity of adenylate cyclase are as follows:

1.   Reduced cellular activities of either Enzyme I or HPr render adenylate cyclase *hypersensitive* to inhibition by any extracellular sugar substrate of the PTS (Figs. 5.5 and 5.6 and Table 5.1).
2.   Inhibition by a particular sugar substrate of the PTS requires that the Enzyme II which recognizes, transports, and phosphorylates that sugar be catalytically active (Saier *et al.,* 1976a).
3.   Mutation of a gene (designated *crrA*, which maps adjacent to the *pts* operon) causes a great reduction in adenylate cyclase activity (Figs. 5.5 and 5.6). *crrA* mutants exhibit reduced activity of the

TABLE 5.1

Effect of Sugars on Cyclic AMP Production by *E. coli* Strains[a]

	Cyclic AMP production in strain	
Sugar added	cpd-454	cpd-454 ptsI109
None	1.4	0.9
Glucose	0.2	0.1
Methyl α-glucoside	1.2	0.2
Mannose	0.9	0.1
Fructose	0.8	0.1
Mannitol	0.1	0.1
Glucitol	0.2	0.2
N-Acetylglucosamine	0.1	0.1
Glucosamine	1.1	0.1
Galactose	0.6	0.4
Melibiose	0.9	0.7
Xylose	0.5	0.7

[a] Cells were grown overnight at 37°C in Medium 63 containing 1% sodium pyruvate and 20 μg of methionine per milliliter, harvested, washed with salts medium, and used to inoculate 10 ml of medium containing 0.5% sodium pyruvate and 20 μg of methionine per ml in 125-ml Erlenmeyer flasks. The flasks were rotated (250 rpm) at 37°C for 3 hours (about 2 generations). Aliquots were removed for cyclic AMP determination, and the sugars indicated in the table were added to a concentration of 0.2% for strain cpd-454 or 0.1% for strain cpd-454 ptsI109. The concentration of cyclic AMP in the culture media at this time was 0.1 μM for both strains. The flasks were then rotated (250 rpm) at 37°C for an additional 3 hours. Aliquots were subsequently removed, and total cyclic AMP was again determined. Values represent the total cyclic AMP produced (nmoles/ml).

glucose Enzyme III which is encoded by the *crrA* gene (see Chapter 4, Section C), implicating this protein in the regulatory process.

4.   Specific regulatory mutations in the *cya* gene, which encodes adenylate cyclase, have been postulated to release adenylate cyclase from PTS-mediated control. Point mutants exhibiting the expected properties have not yet been characterized, but cloned, truncated *cya* genes which do not encode the C-terminus of the protein appear to retain catalytic activity while losing sensitivity to regulation (Roy *et al.*, 1983b; J. Y. J. Wang, personal communication).

5.   Growth of bacteria in the presence of cyclic AMP, incubation of cells in the presence of chloramphenicol, or incubation of cells at 37°C in the absence of a carbon source results in the rapid loss of *in vivo* adenylate cyclase activity (Saier *et al.*, 1982b).

Based largely on these results, a regulatory mechanism involving direct activation of adenylate cyclase by the phosphorylated form of a cytoplasmic regulatory protein was proposed as shown in Fig. 5.7 (Saier and Feucht, 1975; Peterkofsky *et al.*, 1975). While Saier and Feucht proposed that this regulatory protein was Enzyme III^{glc} of the PTS (encoded by the *crrA* gene), Peterkofsky *et al.* suggested that the phosphorylated activator protein was phospho-Enzyme I (Peterkofsky and Gazdar, 1974; Harwood and Peterkofsky, 1975; Peterkofsky *et al.*, 1975; Harwood *et al.*, 1976). These models for adenylate cyclase control were proposed on the basis of genetic and physiological experiments conducted either with intact bacterial cells or with cells treated with toluene (Harwood and Peterkofsky, 1975; Feucht and Saier, 1980). Definitive biochemical studies of the regulatory mechanism have been hampered by the lack of an *in vitro* assay. PTS-mediated regulation of adenylate cyclase has not yet been demonstrated in a cell-free system.

Recently, the regulation of adenylate cyclase was examined in an *E. coli* strain which greatly overproduces the lactose permease due

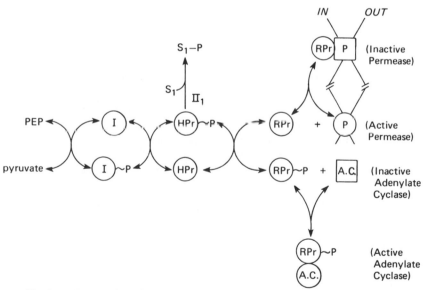

Fig. 5.7. Proposed mechanism for the coordinate regulation of carbohydrate permeases and adenylate cyclase in *E. coli* and *S. typhimurium*. Abbreviations: P, permease; A. C., adenylate cyclase; RPr, regulatory protein, the Enzyme III^{glc}. Other abbreviations are as shown in Fig. 3.1. [From Saier (1977), with permission.]

to the presence of a multicopy plasmid carrying the *lacY* gene (Saier *et al.*, 1983). Employing membranes isolated from this strain, it had been shown that the lactose permease protein binds the free form, but not the phosphorylated form of Enzyme III^{glc} in a process which is dependent on the binding of a galactoside substrate to the permease (see Section C of Chapter 4). Further, the addition of a galactoside substrate of the *lac* permease to intact cells relieved methyl α-glucoside-promoted inhibition of glycerol uptake. The relief of PTS-mediated inhibition of glycerol uptake by increasing concentrations of thiodigalactoside (TDG) was attributed to binding of the allosteric regulatory protein, Enzyme III^{glc}, to the *lac* permease so that it was no longer available in the cytoplasm to bind to the glycerol permease. This interpretation is discussed in Chapter 4, Section C.

If phospho-Enzyme III^{glc} activates adenylate cyclase, and dissociation of Enzyme III^{glc} from the *lac* permease is slow relative to Enzyme III^{glc} phosphorylation, then increasing concentrations of TDG might be expected to inhibit adenylate cyclase since free Enzyme III^{glc} would not be available for phosphorylation by the PTS. On the other hand, if phosphorylation rather than binding of Enzyme III^{glc} to the permease were rate limiting, then TDG should exert no effect. If TDG uptake, accompanied by proton influx, should dissipate the proton motive force significantly, as discussed in Section A of this chapter, inhibition of adenylate cyclase would result upon addition of TDG.

In contrast to these predictions, addition of TDG to *lac* permease-induced cells *relieved* the inhibition of adenylate cyclase caused by methyl α-glucoside addition (Saier *et al.*, 1983). In further studies, qualitatively similar results were obtained after growth of the *lac* permease–overproducing strain under several different conditions (M. J. Novotny and M. H. Saier, Jr., unpublished results). A simple interpretation of these results would be that adenylate cyclase is regulated by a negative control mechanism by free Enzyme III^{glc} rather than by a positive mechanism involving phospho-Enzyme III^{glc} as proposed previously. Alternatively, additional unidentified proteins may be involved in PTS-mediated regulation of adenylate cyclase, or possibly other regulatory interactions have complicated the results.

Relevant to this suggestion, a number of publications provide indirect evidence suggesting that the cyclic AMP receptor protein (CRP) as well as RNA polymerase may either play a role in the regulation of adenylate cyclase or be regulated by the phosphotransferase system (Daniel, 1984; Dobrogosz *et al.*, 1983; Glesyna *et al.*, 1983; Grossman *et al.*, 1984). Dobrogosz *et al.* and Glesyna *et al.* suggested that *pts* mutations influence the functioning of the cyclic AMP receptor protein rather than (or in addition to) that of adenylate cyclase. By contrast, Grossman *et al.* suggested that RNA polymerase and the proteins of the PTS exert their effects primarily on adenylate cyclase.

Daniel (1984) provided evidence that Enzyme III^{glc} (in the presence or absence of Enzyme I) stimulates cyclic AMP synthesis in an *E. coli* mutant which is deleted for the *crp* gene and therefore lacks the cyclic AMP receptor protein (Daniel, 1984). Because all of these investigators restricted their studies to *in vivo* conditions using multiply defective mutant strains, definite conclusions regarding the molecular basis for their observations cannot be drawn at this time. Direct and indirect effects must be distinguished and identified employing *in vitro* assays.

In re-evaluating the model shown in Fig. 5.7, there were originally three lines of evidence suggesting activation by a phosphorylated PTS protein:

1. Phosphoenolpyruvate activated adenylate cyclase activity in toluenized cells (Harwood and Peterkofsky, 1975; Peterkofsky and Gazdar, 1975).
2. Adenylate cyclase activity is very low when measured *in vitro* compared with the *in vivo* activity (Feucht and Saier, 1980).
3. *crrA* mutants of *S. typhimurium* and in most genetic backgrounds of *E. coli* showed low activity (Saier and Feucht, 1975; Feucht and Saier, 1980).

The discrepancy between these observations and the relief of inhibition by TDG in the *lac* permease overproducing strain can be resolved if it is assumed that

1. Adenylate cyclase has low inherent activity.
2. The enzyme is activated by an unidentified protein which not only activates adenylate cyclase, but also renders it sensitive to PTS-mediated inhibition. A phosphorylated form of this protein might conceivably activate adenylate cyclase.
3. It is the binding of free Enzyme III^{glc} to the adenylate cy-

clase–activator protein complex which ultimately inhibits cyclic AMP synthesis. If this new model is correct, it must be further assumed that in *S. typhimurium* and *E. coli*, the *crrA* mutation *prevents* activation of adenylate cyclase by the activator protein.

Still another provocative observation deals with the recent suggestion that Enzyme III^{glc} can be self-phosphorylated at the expense of ATP (Deutscher and Saier, 1983). While the physiological significance of this observation is not yet known, it may be important with respect to the mechanism of adenylate cyclase activation.

The characterization of *crrA* (*crr*) mutants (Saier and Roseman, 1972, 1976a) led to the identification of Enzyme III^{glc} as the central regulatory protein mediating regulation of non-PTS permease function (See Chapter 4, Section C). The *crrA* gene was shown to code for Enzyme III^{glc} and to map adjacent to the *pts* operon in *Salmonella* with the gene order *cysK–ptsH–ptsI–crrA* (Cordaro and Roseman, 1972). More than 90% of all *crrA* mutants exhibited low adenylate cyclase activity *in vivo* and were resistant to PTS-mediated inhibition of non-PTS sugar uptake (Feucht and Saier, 1980). In two different genetic backgrounds of *E. coli* (Crooke's strain and AB257^{PC-1}) *crrA* mutants exhibited the same properties as in *Salmonella* with respect to transport regulation and adenylate cyclase function (see Figs. 5.4 and 5.5). In a third genetic background (that of *E. coli* strain 1100), *crrA* mutants had similar properties except that adenylate cyclase, though partially resistant to inhibition by methyl α-glucoside, exhibited appreciable activity (Feucht and Saier, 1980). These studies were conducted with strains that (a) expressed the glucose Enzyme II inducibly, (b) were wild-type for the expression of the mannose Enzyme II, (c) were deficient in Enzyme I or HPr, and (d) were usually lacking cyclic AMP phosphodiesterase.

Kornberg and his associates have claimed to have identified two genes in *E. coli* which map close to the *pts* operon and influence different aspects of carbohydrate transport and adenylate cyclase regulation (Britton *et al.*, 1983; Parra *et al.*, 1983). One of these genes, termed *iex* (for *i*nducer *ex*clusion), when defective, yielded strains which were partially resistant to inhibition of non-PTS sugar (lactose or maltose) utilization by 5-thioglucose or glucose but showed normal adenylate cyclase activity. The other gene, termed *gsr* (for *g*lucose-*s*pecific *r*esistance) yielded strains which were fully resistant to 5-thioglucose inhibition of lactose and maltose utilization, and adenylate cyclase activity was depressed. A *gsr* mutant was shown to be deficient for Enzyme III^{glc} activity, but a biochemical defect associated with the *iex* mutation was not identified. While *gsr* mutants were reported to take up glucose and methyl α-glucoside poorly, *iex* mutants were unimpaired with respect to this activity. The suggested gene order was *gsr–ptsI–(ptsH, iex)*.

In attempting to explain the differences in the results reported by the Kornberg group and the other laboratories involved in the study of PTS-mediated regulation, the following facts should be kept in mind: First, the genetic backgrounds used to characterize the *gsr* and *iex* mutant alleles varied (Parra *et al.*, 1983), but differed from those used by the other laboratories in studies describing the characterization of *crrA* mutant alleles. The former strains (a) expressed the glucose Enzyme II constitutively, (b) did not express the mannose Enzyme II, (c) were either wild-type for the *pts* operon or contained a temperature-sensitive *ptsI* gene and were (probably) studied at the permissive temperature, and (d) contained wild-type levels of cyclic AMP phosphodiesterase activity. Second, different inhibitory sugars were usually used (5-thioglucose or glucose by the Kornberg group, methyl α-glucoside and other PTS sugars in the other laboratories). Third, growth or sugar utilization over an extended period of time was studied by Kornberg and his co-workers, whereas other laboratories usually examined uptake of [^{14}C]sugars over a short time interval. Finally, the bacterial strains were grown on different carbon sources in the different studies, and different conditions of adenylate cyclase assay were employed.

Based on these differences in strains, growth conditions, and assay conditions, it is not surprising that different results were obtained. *gsr* mutants are probably *crrA* mutants based on (a) the reported map positions of the genes, (b) a deficiency of Enzyme III^{glc} activity reported for the *gsr* mutants, (c) depression in the level of adenylate cyclase activity in *gsr* mutants measured *in vivo* relative to the parental strain, and (d) resistance of non-PTS permeases in *gsr* mutants to inhibition by nonmetabolizable glucose analogues. Inhibition of non-PTS transport activities by metabolizable PTS sugars such as *N*-acetylglucosamine in *gsr* mutants may be attributable to inhibition by intracellular sugar phosphates and other metabolites (see Chapter 4, Section B). Unidentified strain differences may account for some of the differences reported.

Nelson *et al.* (1984) have attempted to resolve the discrepancies previously noted between the Kornberg group and other researchers in the field. In agreement with the aforementioned conclusions, these investigators suggest that *gsr* mutants are, in fact, *crr* (*crrA*) mutants. Moreover, an *iex* mutant of *E. coli*, isolated and characterized by the Kornberg group, was shown to possess an altered Enzyme III[glc] with (1) increased thermolability, (2) decreased activity in the phosphotransferase assay, and (3) an inability to bind to the lactose permease. Introduction of a plasmid carrying a wild-type *crr*[+] allele into the *iex* strain restored the phenotype to that of the wild-type parental strain. The authors therefore concluded that *iex* is allelic with *crr*. That is, the *iex* mutation gives rise to a structurally altered Enzyme III[glc] with decreased affinity for the target permeases, but partial retention of PTS function and adenylate cyclase regulatory function. Although this conclusion is at odds with the genetic mapping data suggesting that *gsr* (*crrA*) and *iex* map on opposite sides of the *ptsI* gene (Britton *et al.,* 1983; Para *et al.,* 1983), it appears to adequately explain all other results. The results are consistent with the conclusion that Enzyme III[glc] plays a primary role in the regulation of both adenylate cyclase and the target permeases, and that *iex*, *gsr*, and *crrA* are allelic. The *iex* and *gsr* designations should therefore be abandoned.

6

Involvement of Protein Kinases in the Regulation of Carbohydrate Transport and Metabolism

In his classic review dealing with the repression of carbohydrate catabolic enzyme synthesis, Magasanik (1970) defined three constituent phenomena: transient repression, catabolite repression, and inducer exclusion (Paigen, 1966). While transient and catabolite repression have both been reported to result from reduced levels of cyclic AMP and the mechanisms for the control of adenylate cyclase discussed in Chapter 5 are probably at least in part responsible, the molecular details, particularly of transient repression, are still poorly understood. Inducer exclusion mechanisms are far better defined as discussed in Chapter 4.

In recent years, an additional regulatory phenomenon has been defined: that of inducer expulsion (Reizer and Panos, 1980). When a potential inducer such as an intracellular sugar phosphate produced by the action of the phosphotransferase system accumulates and when subsequently an efficient energy source is provided, the sugar phosphate is rapidly dephosphorylated, and the sugar exits from the cell. This phenomenon was first described in *E. coli* in the late 1960s and early 1970s by Kepes and his co-workers and was recognized much later in gram-positive bacteria. Evidence for an energy dependence of the process has been presented in both types of organisms, and very recently, an involvement of ATP-dependent protein kinases has been suggested. In this chapter, we shall review some of the early and more recent reports dealing with this phenomenon and shall discuss the possible significance of protein kinases to the regulation of carbohydrate transport and metabolism.

A. FUTILE CYCLES OF SUGAR UPTAKE AND EFFLUX IN
E. COLI AND *S. LACTIS*

In 1962, before the discovery of the phosphotransferase system, Hoffee and Englesberg examined the effects of various energy sources and inhibitors on the kinetics of methyl α-glucoside accumulation in *S. typhimurium* and compared the results with those obtained for β-galactoside uptake via the lactose permease in *E. coli*. They showed that while respiratory activity enhanced the level of β-galactoside uptake, it depressed the extent of accumulation of methyl α-glucoside (Hoffee and Englesberg, 1962). Because anaerobic sugar fermentation had the reciprocal effect, two distinct mechanisms of uptake were proposed. On the basis of these and other results, they suggested that a chemical energy source such as ATP is required not only for methyl α-glucoside uptake, but also for the exit reaction.

Nine years later, in 1971, Haguenauer and Kepes reported the results of kinetic studies on the "cycle of renewal" of intracellular methyl α-glucoside-phosphate accumulated by the glucose phosphotransferase in *E. coli*. The effects of energy inhibitors were studied on the three distinct reactions which comprised the cycle: (a) sugar uptake, coupled to phosphorylation, (b) intracellular dephosphorylation of the cytoplasmic sugar phosphate, and (c) efflux of the free sugar from the cell. Their results foreshadowed several fascinating phenomena, which are still under intensive investigation. For example, they showed that energy inhibitors such as azide and dinitrophenol *stimulated* methyl α-glucoside uptake by reducing the K_m for uptake, a result which we now know to be due to the inhibitory effect of the membrane potential on the activity of the glucose Enzyme II (see Chapter 4, Section A).

Under normal conditions, in the absence of an energy inhibitor, intracellular concentrations of free and phosphorylated methyl α-glucoside rapidly attained steady state levels (in about 15 minutes at 25°C), after which time the rates of uptake, dephosphorylation, and efflux of free sugar were all equal. Subsequent addition of 50 mM NaF (an inhibitor of glycolysis) together with 40 mM NaN$_3$ (an inhibitor of electron transport, which at high concentrations can also block the functioning of the proton-translocating ATPase and dissipate the membrane potential) resulted in a transient increase in the intracellular sugar phosphate pool. Net efflux of the free cytoplasmic sugar continued unabated. The inhibitors apparently blocked uptake (presumably due to depletion of the intracellular pool of phosphoenolpyruvate) but more importantly, they blocked sugar phosphate dephosphorylation without inhibiting efflux of the free sugar. Because dephosphorylation and efflux were not blocked by treatment with N-ethylmaleimide under conditions which resulted in inactivation of the glucose Enzyme II, it was concluded that this enzyme did not play a role in the expulsion process. This last conclusion may be incorrect as discussed in Section B of this chapter.

The authors assumed that the energy source for activation of the sugar phosphate phosphatase was phosphoenolpyruvate primarily because of the role of this compound in driving sugar uptake. However, in retrospect, ATP appears to be a more probable phosphoryl donor, and activation of cytoplasmic sugar phosphate phosphatase by a protein kinase is proposed. Known sugar phosphate phosphatases are localized to the periplasmic space (Rephaeli et al., 1980; Kier et al., 1977a,b), and no known enzyme which hydrolyzes cytoplasmic sugar phosphates has yet been identified. Since mutant enteric bacteria lacking the hexose-phosphate phosphatase are not known and should be difficult to select, a possible solution to this problem would be to release the soluble periplasmic enzymes by spheroplasting or by osmotic shock and then to seek the identification of an ATP-activatable methyl α-glucoside phosphate phosphatase. The identification of this enzyme cascade remains future work for the microbial biochemist.

A phosphoenolpyruvate-dependent futile cycle has also been identified in *Streptococcus lactis* (Thompson and Chassey, 1982). In this organism, energy-starved cells accumulate high concentrations of the potential phosphoryl donors: phosphoenolpyruvate, 2-phosphoglycerate, and 3-phosphoglycerate. The starved cells are therefore "triggered" for the uptake of sugar substrates of the PTS. If a nonmetabolizable glucose analogue is added to a starved cell suspension, the sugar is rapidly accumulated to a concentration of nearly 100 mM, until the glycolytic intermediates are exhausted. Subsequently, slow intracellular dephosphorylation of the sugar phosphate occurs, and the free sugar "leaks" out of the cell. This cycle utilizes so much energy in energy proficient cells that growth stasis can result (Thompson and Chassey, 1982).

Addition of extracellular nonradioactive 2-deoxyglucose or the glycolytic inhibitor, iodoacetate, resulted in slow loss of the intracellularly accumulated 2-deoxy[^{14}C]glucose 6-phosphate. However, when a metabolizable sugar such as glucose was added, the rates of dephosphorylation and efflux of the accumulated [^{14}C]sugar phosphate increased at least 10-fold. Phosphatase action was the rate-limiting step. The results can be explained if it is assumed that ATP, possibly in conjunction with a glycolytic intermediate, activates the phosphatase by a protein kinase-mediated mechanism. This result appears to be analogous to that reported for *E. coli* by Haguenauer and Kepes (1971).

B. PHYSIOLOGICAL CHARACTERIZATION OF INDUCER EXPULSION

In gram-positive bacteria, the process of inducer expulsion has been studied in detail. The first careful study of this phenomenon was published by Reizer and Panos in 1980. They showed that *Streptococcus pyogenes* utilized the PTS for the uptake of lactose and its nonmetabolizable analogues and that the galactoside phosphates accumulated in the cytoplasm. When a nonmetabolizable [^{14}C]galactoside was employed, the intracellular pool of sugar phosphate was stable and did not turn over. If glucose, mannose, or glucosamine was subsequently added to the culture fluid, these metabolizable hexoses induced rapid expulsion of the radioactivity from the galactoside-phosphate pool, and the free sugar was released into the external medium. The half-time for expulsion of 30 mM [^{14}C]TMG was about 15 seconds, a time much shorter that the half-time for maximal accumulation of the sugar phosphate under similar conditions. Nonmetabolizable glucose analogues such as 6-deoxyglucose and 2-deoxyglucose were much less effective in inducing efflux than were the metabolizable sugars. Moreover, prepoisoning the cells with fluoride or arsenate prevented glucose-induced expulsion of the galactoside. Several lines of evidence argued against a vectorial transphosphorylation process. It was suggested that the rapid expulsion process was relevant to induction of expression of the galactoside catabolic regulon in gram-positive bacteria where the natural inducer appears to be galactose 6-phosphate (Reizer and Panos, 1980; Simoni and Roseman, 1973). The authors coined the term "inducer expulsion," to be contrasted with "inducer exclusion," which involves inhibition of inducer uptake (Chapter 4).

The same process was studied in greater detail by Thompson and Saier (1981) employing *Streptococcus lactis* strain ML$_3$. When these bacteria were starved for an energy source, they accumulated cytoplasmic 2-phosphoglycerate, 3-phosphoglycerate and phosphoenolpyruvate to a net internal concentration of about 30 mM. Further metabolism of these compounds (which were in equilibrium with each other) did not occur because in *S. lactis* and various other streptococci, the next catabolic enzyme in the pathway, pyruvate kinase (which converts phosphoenolpyruvate and ADP into ATP and pyruvate), is allosterically activated by fructose 1,6-diphosphate. In the absence of this activator, the enzyme is essentially inactive. When the starved *S. lactis* cells were exposed to a radioactive sugar substrate of the PTS, the sugar was rapidly accumulated intracellularly as the phosphate ester. The final concentration of sugar phosphate approached that of the initial phosphoenolpyruvate plus phosphoglycerate pool. An example of this behavior is shown in Fig. 6.1.

If the intracellular sugar phosphate was TMG 6-phosphate, subsequent addition of glucose, mannose, or lactose induced rapid expulsion of the galactoside from the cell (Fig. 6.1). While maltose alone was inactive in promoting efflux, maltose plus arginine was as effec-

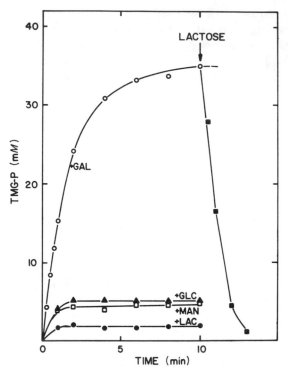

Fig. 6.1. Exclusion and expulsion of thiomethyl β-galactoside (TMG) from *Strepto-coccus lactis*. *S. lactis* strain ML₃ was grown, washed, and starved for energy as described by Thompson and Saier (1981). The washed cells were exposed to 0.4 mM [¹⁴C]thiomethyl β-galactoside in the presence of galactose (○), glucose (▲), mannose (□), or lactose (●). The metabolizable sugars were present at a concentration of 1 mM. As can be seen, the metabolizable sugar substrates of the phosphotransferase system, but not galactose, which is taken up by an active galactose permease, excluded TMG from the cell. After a 10 minute incubation in the presence of galactose, lactose (1 mM) was added which elicited rapid efflux of the preaccumulated TMG (closed squares). [J. Thompson and M. H. Saier, Jr., unpublished results.]

tive as glucose. The arginine deiminase pathway was presumed to stimulate efflux in the presence of maltose by production of ATP. As concluded by Reizer and Panos, it appeared that both entry and metabolism of the hexose were required to elicit galactoside efflux since a mutant lacking the glucose (mannose) Enzyme II was resistant to hexose-promoted galactoside expulsion. It was shown that intracellular dephosphorylation preceded efflux of the free sugar.

Thus, in agreement with the suggestion of Reizer and Panos, a vectorial phosphorylation reaction could not have been responsible for efflux of the intracellularly formed free galactoside.

Results reported by Reizer and Panos (1980) and Thompson and Saier (1981) revealed that the presence of a sugar, such as glucose, mannose, or 2-deoxyglucose, also prevented accumulation of the galactoside as shown in Fig. 6.1. Although this effect was assumed to be due exclusively to competition for phospho-HPr, an additional mechanism of galactoside exclusion involving protein kinase-catalyzed phosphorylation of a seryl residue in HPr is now believed to be operative (see Section C of this chapter).

The results presented above suggested that ATP as well as a glycolytic intermediate might be required for intracellular dephosphorylation of the galactoside phosphate. A search was therefore made for an ATP-activated sugar-phosphate phosphatase (Thompson and Saier, 1981). Cells were permeabilized with toluene, and the hydrolysis of TMG-6-P was studied in the presence of various compounds. TMG-6-P hydrolysis was stimulated two- to threefold by ATP, but was strongly inhibited by fluoride. These observations suggested the possible involvement of a protein kinase in activation of the sugar phosphate phosphatase, and that this protein kinase might be activated by a glycolytic intermediate.

A subsequent report (Reizer et al., 1983) served to confirm the two-step mechanism and to eliminate the possibility of an involvement of a vectorial transphosphorylation process. When the cells were allowed to accumulate [^{14}C]TMG-P, washed free of [^{14}C]TMG, exposed to a high concentration of extracellular [^{12}C]TMG or IPTG, and then exposed to glucose, net efflux of TMG was strongly inhibited relative to the control in which the nonradioactive galactoside was absent (Fig. 6.2). Further analyses revealed that the [^{12}C]galactosides inhibited efflux without inhibiting intracellular dephosphorylation of the TMG-phosphate as shown in Fig. 6.2. Genetic loss of the lactose Enzyme II prevented glucose-promoted efflux of TMG from the cells (Reizer and Saier, 1983). Moreover, specific competitive inhibitors of the Enzyme II[lac], but not competitive inhibitors of the galactose permease, greatly hindered efflux. On the basis of these results, it was suggested that the route of galactoside efflux was the lactose Enzyme II (Reizer and Saier, 1983). Table 6.1 summarizes the physiological characteristics of the inducer expulsion process.

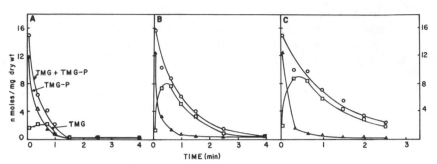

Fig. 6.2. Intracellular content of TMG, TMG-P, and TMG plus TMG-P during expulsion. *S. pyogenes* cells were preloaded with [^{14}C]TMG-P for 7.5 min at 37°C and collected by centrifugation. The cell pellets were suspended in buffered medium containing (A) NaF (10 mM) and arginine (5 mM); (B) NaF (10 mM), arginine (5 mM), and unlabeled TMG (100 mM); or (C) NaF (10 mM), arginine (5 mM), and IPTG (100 mM). After preincubation (5 min) at 37°C, expulsion was elicited by addition of 10 mM glucose. Duplicate samples were removed at the indicated intervals for determination of (○) total intracellular radioactivity, (□) free TMG, and (△) TMG-P. [From Reizer (1983), with permission.]

Recent studies with *Lactobacillus casei* have shown that pentitol phosphates (ribitol-5-P, and xylitol-5-P) are accumulated by pentitol-specific phosphotransferases (London and Chace, 1977, 1979; London and Hausman, 1982). One natural isolate of this organism accumulates high cytoplasmic concentrations of xylitol-5-P but cannot metabolize this compound further because xylitol-5-P dehydro-

TABLE 6.1

Summary of Physiological and Biochemical Studies Aimed at Characterizing the Inducer Expulsion Phenomenon in *Streptococcus pyogenes*

A. Physiological characteristics
    1. Inducer expulsion is a two-step process:
        (a) Sugar-P hydrolysis
        (b) Sugar efflux
    2. Sugar-P hydrolysis requires the simultaneous presence of ATP and a glycolytic intermediate
    3. Sugar efflux is catalyzed by the lactose Enzyme II of the PTS
B. Biochemical characteristics
    1. A protein kinase phosphorylates a (regulatory) seryl residue in HPr
    2. HPr(ser) phosphorylation is subject to positive control by carbohydrate metabolic intermediates
    3. An HPr(ser)P phosphatase reverses the kinase reaction

genase is lacking. When this strain is allowed to accumulate [^{14}C]xylitol-5-P and is subsequently exposed to glucose, the xylitol-5-P is rapidly lost from the cytoplasmic pool, and the radioactive pentitol appears in the extracellular medium. Because the PTS provides the only transport pathways for pentitols in this organism, it must be considered that the Enzymes II mediate the expulsion of the free pentitol (J. London, personal communication). This conclusion is in agreement with that of Reizer and Saier (1983) for the extrusion of TMG from *Streptococcus pyogenes*. It is worthy of note that Haguenauer and Kepes (1971) concluded that the glucose Enzyme II was not responsible for the energy-dependent efflux of methyl α-glucoside in *E. coli*. Another permease was presumed to catalyze efflux of the sugar in this case. The conclusion was based on the fact that *N*-ethylmaleimide inhibited methyl α-glucoside uptake, but not its efflux. Possibly the Enzyme II mediates both processes in *E. coli* as in the gram-positive bacteria, but the essential sulfhydryl group of the enzyme is not required for efflux.

In agreement with this suggestion, *N*-ethylmaleimide hinders but does not block TMG efflux from *S. pyogenes* under conditions in which TMG uptake is completely inhibited (J. Reizer and M. H. Saier, unpublished results). This observation, together with the rapid rate of efflux relative to uptake, suggests that while uptake involves catalytic turnover of the transport system, efflux may involve passage through a channel within the same permease protein, the Enzyme II specific for that sugar. This observation may have tremendous implications with respect to the mechanism of Enzyme II-catalyzed sugar transport (see Chapter 3, Section E).

Further *in vivo* studies have substantiated the conclusion of Thompson and Saier that ATP and a glycolytic intermediate are required for intracellular hydrolysis of the galactoside phosphate (Reizer *et al.*, 1983). Consequently, an *in vivo* search for protein phosphorylation was undertaken. *S. pyogenes* cells were incubated with [^{32}P]inorganic phosphate under a variety of conditions, including those which promoted TMG-P expulsion. Subsequent SDS-gel electrophoresis revealed the presence of a single low molecular weight [^{32}P]phosphoprotein which was only present under the conditions of expulsion (Reizer *et al.*, 1983). The *in vitro* conditions of its phosphorylation and dephosphorylation, the identity of the protein as HPr, and the possible physiological significance of this protein kinase-catalyzed reaction will be discussed in the next section.

C. HPr KINASE AND HPr(ser)P PHOSPHATASE IN
S. PYOGENES

As revealed by the *in vivo* phosphorylation results described in the previous section, a single phosphoprotein of low molecular weight appeared to be generated under the conditions of inducer expulsion. Phosphorylation of this protein with γ-[^{32}P]ATP could be demonstrated in crude extracts of *S. pyogenes*, and the phosphorylation reaction was stimulated by inclusion of 2-phosphoglycerate or glucose 6-phosphate. The former compound was more effective than the latter, and fructose 1,6-diphosphate was reported to exhibit marginal activity. Other glycolytic intermediates were without effect (Deutscher and Saier, 1983).

The [^{32}P]phosphoprotein was purified by DEAE cellulose column chromatography and by gel filtration. Because the phosphoprotein exhibited properties similar to those of HPr, fractions were also assayed for this protein. While a single peak of radioactivity from the Sephadex G-75 column co-eluted with HPr, two peaks of HPr activity eluted from the DEAE cellulose column. The smaller peak coincided with the peak of radioactivity; the larger peak eluted in front of the radioactive one. It was reasoned that phosphorylation of the protein, which should increase its negative charge, caused it to adhere more tightly to the resin. This fact allowed separation of the free protein from the phosphorylated form.

The [^{32}P]phosphoprotein was nearly electrophoretically pure after these two purification steps. It migrated together with HPr in SDS gels, but the phosphoprotein migrated on native gels as a protein with one additional negative charge. The phosphoprotein was both acid and base stable, eliminating the possibility of an acyl or histidyl phosphate. Therefore, HPr was presumed to be phosphorylated on a seryl, threonyl, or tyrosyl residue.

To distinguish between these three possibilities, the phosphoprotein was hydrolyzed in 6 N HCl at 110°C. Electrophoresis of the hydrolysate revealed only [^{32}P]phosphoserine. No phosphothreonine or tyrosine was detected (Deutscher and Saier, 1983). Consequently, it was concluded that a seryl residue in HPr was phosphorylated. Experiments showed that this phosphoryl group could not be transferred to sugar in the sequence of phosphoryl transfer reactions. The phosphoryl moiety of the phosphoseryl residue was evidently not in equilibrium with or interconvertible with the phosphoryl moiety attached to the active histidyl residue of the protein.

A crude extract derived from *S. pyogenes* cells was found to contain a soluble HPr(ser)phosphate phosphatase. It eluted from a Sephadex G-75 column near the void volume with an apparent molecular weight of about 70,000. On DEAE cellulose, it eluted at a high salt concentration (0.55 M NaCl). It lost activity in solutions of high salt content. However, it was observed that the phosphatase activity was strongly inhibited by addition of phosphoenolpyruvate and purified Enzyme I. Because neither Enzyme I nor phosphoenolpyruvate alone inhibited the reaction, it was presumed that phosphorylation of the histidyl residue in P-(ser)HPr inhibited phosphatase activity. The doubly phosphorylated protein appeared to be a poor substrate of the phosphatase.

The HPr kinase was found to be membrane-associated, but exposure to 100 mM phosphate buffer, pH 7.5, for 24 hours released the activity in a soluble form. It eluted from Sephadex G-75 as a small protein (MW = 20,000). Surprisingly, the activity of the partially purified protein was not stimulated by glucose 6-P, and it was only slightly stimulated by 2-phosphoglycerate. It was therefore considered either that the protein normally existed in association with a membrane-bound regulatory subunit or that it had been proteolytically cleaved from its regulatory moiety during the 24-hour incubation in phosphate buffer. The major biochemical observations regarding HPr phosphorylation are summarized in Table 6.1.

More recently the HPr kinase has been extracted under similar conditions but for a shorter period of time (2 hours) in the presence of the protease inhibitor, phenylmethylsulfonyl fluoride (J. Reizer and M. H. Saier, Jr., unpublished results). The kinase extracted in this way eluted from a Sephadex G-75 column near the void volume, suggesting that it had a much larger size. By SDS polyacrylamide gel electrophoresis, the major protein had an apparent subunit molecular weight of 60,000. Moreover, the activity of the partially purified protein was greatly stimulated by fructose 1,6-diphosphate and gluconate-6-P. Glucose-6-P was much less stimulatory. The differences between the results reported originally by Deutscher and Saier (1983), and those obtained by Reizer and Saier have yet to be explained. Possibly proteolysis of a regulatory moiety of the kinase during extended incubation accounts for these differences.

As noted above, HPr kinase activity in crude extracts was absolutely dependent on ATP and one of several carbohydrate metabolic intermediates, and the *in vivo* conditions which resulted in HPr phosphorylation were the same as those which promoted expulsion of TMG from the cell. Since the rate-limiting step for efflux which was activated by ATP was sugar phosphate hydrolysis, it was presumed that phospho(ser)HPr activates a sugar phosphate phosphatase *in vivo* (Fig. 6.3). The possibility of the involvement of a minor (undetected) phosphoprotein which was phosphorylated in parallel with HPr could not be ruled out.

The scheme shown in Fig. 6.3 suggests that phospho(ser)HPr (but not free HPr) functions to activate a cytoplasmic sugar-P phosphatase. The P(ser)HPr is generated by an ATP-dependent HPr kinase

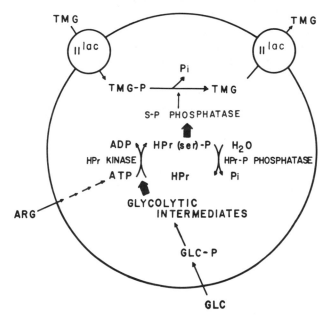

Fig. 6.3. Proposed scheme for the expulsion of TMG-P from streptococcal cells. The scheme shows entry of TMG via the Enzyme II^{lac} of the PTS. Dephosphorylation of cytoplasmic TMG-P is catalyzed by a sugar phosphate phosphatase which is presumed to be stimulated by HPr(ser)P. Free TMG is thought to exit from the cell via the Enzyme II^{lac}. ATP-dependent phosphorylation of a seryl residue in HPr is catalyzed by an HPr kinase which is activated by glycolytic intermediates. While cytoplasmic ATP can be derived from exogenous arginine, glycolytic intermediates result from the uptake and metabolism of glucose. The phosphoryl moiety of HPr(ser)P is hydrolyzed by an HPr-P phosphatase. The specificities of these enzymes for their protein substrates are not known.

which is activated by glycolytic intermediates. It is hydrolyzed by the HPr(ser)P phosphatase. Identification of the sugar-P phosphatase has not been accomplished, but it may be the same as the enzyme purified by Thompson and Chassey (1983). Further experiments will be required to establish the veracity of the scheme proposed in Fig. 6.3.

Recently it has been found that phosphorylation of the seryl residue in HPr alters its reactivity in the phosphotransfer reactions (J. Deutscher, unpublished results). After separation of P(ser)HPr from free HPr by DEAE cellulose chromatography, the two HPr preparations were tested for their sugar phosphorylating activities both in the lactose and in the gluconate phosphotransferase reactions. While the latter reaction was virtually unaffected by phosphorylation of the seryl residue in HPr, the former reaction was strongly inhibited. Preliminary evidence suggested that regulation occurred at the level of Enzyme I–catalyzed phosphorylation of HPr. Enzyme I did not appreciably phosphorylate HPr(ser)P, but complexation of this phosphoprotein with the gluconate Enzyme III stimulated the phosphoryl transfer reaction from phospho(his)Enzyme I to the histidyl residue of P(ser)HPr several orders of magnitude. The gluconate Enzyme III did not stimulate transfer of phosphate from phospho(his)Enzyme I to the histidyl residue of free HPr, and Enzyme III[lac] was much less stimulatory than the gluconate Enzyme III in promoting HPr(ser)P phosphorylation (J. Deutscher, unpublished results). It appears that phosphorylation of a seryl residue in HPr allows complexation with the different Enzymes III to regulate the rate of phosphoryl transfer from Enzyme I to P(ser)HPr. This step presumably controls the rate of sugar phosphorylation and sets up a hierarchy of preferences for the utilization of PTS sugars.

Whether this mechanism (suggested for the gram-positive bacteria) or simple competition for HPr (suggested for the gram-negative bacteria) is the sole or predominant mechanism of regulation for any one organism may depend on the organism under study. The observations described in the preceding paragraph suggest that phosphorylation of the seryl residue in HPr not only stimulates expulsion of galactosides from the cell, its also inhibits uptake and phosphorylation of galactosides via the sequence of PTS-mediated phosphoryl transfer reactions. Thus, in this case a single protein kinase-catalyzed event may be responsible (or partially responsible) for both inducer expulsion and inducer exclusion. The details of these processes may be revealed by further studies in the near future.

A cursory search of other bacterial species has revealed the presence of HPr kinase and P(ser)HPr phosphatase activities in other gram-positive bacteria such as *S. faecalis*, *S. lactis*, *S. aureus*, and *B. subtilis* (J. Reizer, M. J. Novotny, and M. H. Saier, unpublished results). In the gram-negative bacterium, *E. coli*, ATP-dependent phosphorylation of Enzyme III^{glc} as well as HPr was detected (J. Deutscher, J. L. Guan, and M. H. Saier, Jr., unpublished results). However, the ATP-dependent phosphorylation of HPr has not been confirmed in more recent studies (M. J. Novotny, J. Reizer, and M. H. Saier, Jr., unpublished results). The phospho-Enzyme III^{glc} formed from γ-[^{32}P]ATP appeared to be acid and base sensitive. The protein appeared to catalyze self phosphorylation in the absence of a kinase or other proteins of the PTS. Because of the role of this enzyme as a central regulatory protein for the control of inducer exclusion and cyclic AMP synthesis (see Chapters 4 and 5), the physiological significance of this observation may prove to be most intriguing.

Finally, partial separation of proteins within an *S. pyogenes* extract by DEAE cellulose column chromatography allowed detection of additional phosphoproteins after incubation with γ-[^{32}P]ATP (J. Reizer, M. J. Novotny, and M. H. Saier, Jr., unpublished results). The most prominent of these proteins has an apparent molecular weight of 60,000. These findings are of particular interest because 60,000 is the apparent molecular weight of a glucose Enzyme III in gram-positive bacteria (W. Hengstenberg, M. J. Novotny, J. Reizer, and M. H. Saier, Jr., unpublished observation). The same value has been reported for Enzyme I of the PTS, the HPr-kinase, the HPr(ser)P phosphatase discussed above, and the sugar phosphate phosphatase purified by Thompson and Chassey when analyzed by SDS gel electrophoresis. Whether or not these proteins are one and the same or distinct, and their relationships to the phosphorylated protein of this molecular weight remains to be determined.

D. PROPERTIES OF BACTERIAL PROTEIN KINASES IN
S. TYPHIMURIUM

For many years, bacterial protein kinases were thought to be nonexistent. Most attempts to demonstrate the presence of such enzymes were unsuccessful, partly because the substrates employed in such studies were histones and other eukaryotic proteins which had been known to be phosphoryl acceptors of animal cell protein kinases. It now appears that the bacterial protein kinases, in contrast to eukaryotic protein kinases, are highly specific for their target substrate(s). Demonstration of their activities required the use of endogenous bacterial protein substrates under highly specific conditions.

The initial studies revealed that serine and threonine residues in the target proteins were phosphorylated in enteric bacteria (Wang and Koshland, 1978, 1981a,b). The proteins which could be phosphorylated *in vitro* with γ-[^{32}P]ATP were also labeled *in vivo* when cells were incubated with inorganic [^{32}P]phosphate. This result suggested that phosphorylation was not just an artifact of the *in vitro* system, but was a physiologically relevant phenomenon. More recently, the phosphorylation of tyrosyl residues in *E. coli* proteins has been demonstrated (Manai and Cozzone, 1983). In view of these results, it appears that each type of protein kinase found in higher organisms is also present in bacteria. The scope of the physiological significance of this observation has yet to be established.

Two-dimensional gel electrophoretic analysis of the [^{32}P]phosphoproteins, labeled in intact *S. typhimurium* cells, revealed 10 such proteins (Wang and Koshland, 1981a). Their molecular weights varied from 22,000 to 88,000, and their isoelectric points were between 5.4 and 6.6. Proteins phosphorylated to the extent of 200 copies per cell could be detected by the method employed, but some of the proteins were phosphorylated to a much greater extent.

Bacterial proteins which had been shown to be phosphorylated by eukaryotic protein kinases, including RNA polymerase, initiation factor IIb, and ribosomal proteins were not phosphorylated in the *E. coli in vitro* phosphorylation system. Of the proteins phosphorylated *in vitro*, only one has been identified: isocitrate dehydrogenase (MW = 46,000).

In order to estimate the number of protein kinases and protein phosphate phosphatases in the extract, the extract was fractionated using ATP agarose affinity chromatography and gel filtration. Fractions were then assayed for ATP-dependent phosphorylation before and after combination with other fractions. Based on their chromatographic behavior, their substrate specificities and their sensitivities to inhibition (by a small endogenous heat stable inhibitor, various nucleotides and pyrophosphate) four protein kinases were identified. Their characteristics are presented in Table 6.2. It is interesting to note that each kinase phosphorylates different protein substrates with the sole exception of the 45K protein which is phosphorylated by both kinases I and II. Interestingly, kinase II was not active after growth in glucose containing medium, suggesting that it might be subject to catabolite repression. The 45 K protein contains phosphothreonine and could be resolved into two spots on an isoelectric focusing gel. Therefore, it may contain two distinct phosphorylation sites (Wang and Koshland, 1978, 1981a,b).

Enami and Ishihama (1984) have confirmed and extended the findings of Wang and Koshland. They demonstrated that more than 40 distinct protein species were phosphorylated on serine and threonine residues both *in vivo* and *in vitro*. Two protein kinases were purified. One had a molecular weight of 100,000 and catalyzed self-phosphorylation. The other, a dimeric enzyme with a molecular weight of 120,000, catalyzed the phosphorylation of a protein with a molecular weight of 90,000. The functions of these kinases have yet to be defined.

Pulse chase experiments established that many of the substrate phosphoproteins could be dephosphorylated in the crude extract (Wang and Koshland, 1978, 1981a,b). The rates of hydrolysis of the phosphate esters were dependent on the substrate under study. A heat sensitive protein phosphate phosphatase activity was identified, and it was concluded that a second phosphatase must be present in intact cells because the 46K protein (isocitrate dehydrogenase) was reversibly phosphorylated *in vivo* but irreversibly phosphorylated *in vitro* (see next section).

Because cyclic AMP stimulates the activities of several classes of

TABLE 6.2

Characteristics of Four Protein Kinases in *S. typhimurium*

Kinase	Substrates	Inhibitors
I	88K and 45K proteins	Pyrophosphate, phosphate, GTP
II	46K and 45K proteins	AMP, GTP, endogenous heat-stable inhibitor
III	53K protein	AMP, ADP, GTP
IV	63K protein	GTP, AMP, Na$^+$

protein kinases in eukaryotic cells, adenylate cyclase-negative mutants, which lack cyclic AMP, were examined for their phosphorylation patterns (Wang and Koshland, 1981a,b). Most of the major phosphoproteins detected in the wild-type cell were present in the adenylate cyclase-negative mutant. The 46K protein (isocitrate dehydrogenase) was not phosphorylated unless cyclic AMP was added to the growth medium, presumably because an essential protein component is synthesized in response to cyclic AMP. Relevant to this possibility was the fact, noted above, that kinase II, which phosphorylates isocitrate dehydrogenase (Table 6.2), was not active in glucose-grown cells. The *in vitro* activity of kinase II was not stimulated by addition of cyclic AMP (Wang and Koshland, 1981a). In contrast to these observations with *S. typhimurium*, Malloy and Reeves (1983) reported that in an *E. coli* adenylate cyclase deletion mutant the kinase for isocitrate dehydrogenase is not only synthesized but is fully active. Possible differences exist in the regulation of kinase II synthesis in these two organisms.

While the work described in this section clearly establishes the presence of protein kinases in bacteria, some caution must be exercised when identifying ATP-dependent protein kinase reactions. Bacterial extracts have been shown to catalyze rapid [^{32}P]ATP–phosphoenolpyruvate exchange, and phosphoenolpyruvate effectively phosphorylates the proteins of the PTS as well as some additional enzymes (E. B. Waygood, unpublished observations). Consequently, it should not be concluded that ATP is the immediate phosphoryl donor unless nonradioactive phosphoenolpyruvate does not diminish the protein labeling from [^{32}P]ATP and unless the phosphorylated amino acyl residue in the protein is characterized as a seryl, threonyl, or tyrosyl residue.

E. ISOCITRATE DEHYDROGENASE PHOSPHORYLATION AND THE REGULATION OF CARBON METABOLISM

Isocitrate dehydrogenase is a key enzyme which determines the pathway of carbon flow (Garnak and Reeves, 1979; Wang and Koshland, 1982). Under normal aerobic conditions when bacteria are utilizing a carbohydrate for growth, isocitrate dehydrogenase is active, and the Krebs cycle functions for the catabolism of acetate to carbon dioxide. When cells are forced to utilize acetate or ethanol as the sole source of carbon, isocitrate dehydrogenase becomes inactivated, and the glyoxylate shunt (Fig. 6.4) is activated for anabolic purposes. In *Salmonella,* the inactivation of isocitrate dehydrogenase (IDH) occurs by a phosphorylation mechanism catalyzed by a protein kinase which is activated *in vivo* by acetate, ethanol, and nonmetabolizable glucose analogues (Wang and Koshland, 1982).

Employing an *in vivo* ^{32}P–pulse labeling procedure, the response of IDH to shifts in carbon sources can be followed. Maximal labeling of the enzyme in washed acetate-grown cells took about 5 minutes following addition of acetate, a time course which closely followed inactivation of the enzyme. Similar inactivation was observed when ethanol, methyl α-glucoside, or 2-deoxyglucose was added to the cell suspension. Pyruvate and glucose did not enhance phosphorylation although they appeared to accelerate turnover of the label. None of these compounds activated phosphorylation *in vitro* (Wang and Koshland, 1982). Phosphorylation was not observed in glucose-

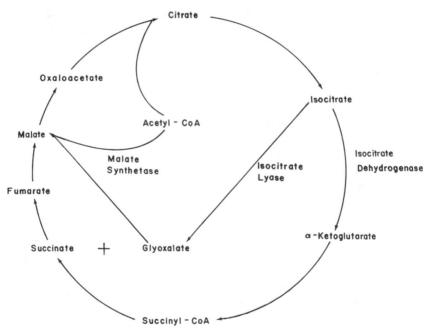

Fig. 6.4. Abbreviated version of the Krebs cycle (outer circle) and the glyoxylate shunt (center). Only the enzymes which are relevant to the activity and control of the glyoxylate shunt are shown.

grown cells, apparently because synthesis of the kinase was poorly induced or repressed. It is interesting to note that isocitrate lyase and malate synthase (Fig. 6.4), two enzymes specific for the glyoxylate shunt, were also induced by growth in acetate and were present in low amounts when cells were grown in glucose-containing medium. Induction of both activities and of the kinase took about 1 hour following transfer to acetate-containing medium, and induction could be blocked by chloramphenicol. Since transfer of acetate-grown cells to glucose medium did not eliminate kinase activity, it was concluded that the enyzme is stable and not inactivated by the glucose metabolites. Rather, its loss during growth in glucose medium was attributed solely to dilution under conditions which resulted in low synthetic rates. Possibly, the kinase and the two glyoxylate cycle enzymes are coordinately regulated because they are encoded within a single operon or regulon. It seems that conditions which result in synthesis of the glyoxylate cycle enzymes simultaneously inactivate isocitrate dehydrogenase, causing the flow of carbon to be shifted from the Krebs cycle into the shunt. This system serves as the first, and at present, the only well-characterized example of a metabolic pathway in bacteria regulated by protein phosphorylation.

The protein kinase responsible for the phosphorylation of isocitrate dehydrogenase has been purified to homogeneity in a three-step procedure involving (1) ammonium sulfate fractionation, (2) ion exchange on DEAE Sephacel, and (3) affinity chromatography using IDH immobilized on Sepharose 4B (La Porte and Koshland, 1982). The kinase activity was found to co-elute from the ion exchange and affinity chromatography columns together with an ATP-dependent IDH-phosphate phosphatase activity. These two activities also co-eluted following gel filtration. Moreover, they bound to the IDH affinity column much more tightly in the presence of ATP (or the ATP analog, adenylylimido diphosphate) than in its absence, and ATP removal resulted in co-elution of the two activities. Since polyacrylamide gel electrophoresis in the presence of sodium dodecyl sulfate revealed a single polypeptide chain, it would appear that the two activities are associated with this one protein.

A kinetic analysis of phosphatase activation by adenine nucleotides showed that ATP activated the enzyme with a K_a of 100 μM while ADP activated less effectively with a K_a of 400 μM. Addition of an ATP regenerating system did not inhibit phosphatase activity with either compound. Further, [^{32}P]ATP was not formed when [^{32}P]phospho IDH was hydrolyzed in the presence of either ADP or ATP. The products of the hydrolytic reaction were therefore presumed to be free IDH and inorganic phosphate. These results lead to the probability that the function of ATP (or ADP) is to allosterically activate IDH–phosphate phosphatase.

The results described above suggest that a single enzyme complex (possibly a single polypeptide chain) catalyzes both the forward and reverse reactions responsible for the reversible covalent inactivation of IDH. Two other enzyme complexes have been shown to similarly catalyze both synthesis and hydrolysis of phosphate esters. One of these enzymes is responsible for the adenylylation and deadenylylation of glutamine synthetase in *E. coli* (Caban and Ginsburg, 1976), and the other enzyme complex (from rat liver) catalyzes both the synthesis and hydrolysis of fructose 2,6-diphosphate from and to fructose-6-P (El-Maghrabi *et al.*, 1982). It is worth noting that all three enzyme complexes function to regulate intermediary metabolism. The occurrence in a single enzyme, or complex of enzymes, catalyzing both the forward and reverse reactions may insure, or at least facilitate, coordination of the activities of these two opposing functions. Such an occurrence may be restricted to enzymes which play essential regulatory roles in controlling cellular metabolic processes.

F. EVOLUTION OF CYCLIC AMP AND PROTEIN KINASES AS REGULATORY AGENTS FOR THE CONTROL OF CARBON AND ENERGY METABOLISM

Protein kinases have long been recognized as important regulatory catalysts for the control of cellular catabolic, anabolic, and differentiative processes in higher organisms (Krebs and Beavo, 1979). For example, cyclic AMP-dependent protein kinases regulate the rates of glycogen, lipid, and amino acid catabolism in animals. More recent studies, summarized in the preceding sections of this chapter, reveal that metabolite-activated protein kinases regulate carbohydrate uptake and metabolism in bacteria (Fig. 6.5). The two target bacterial proteins of protein kinases so far identified are isocitrate dehydrogenase in enteric bacteria and HPr of the PTS in gram-positive bacteria. Phosphorylation of the former protein presumably

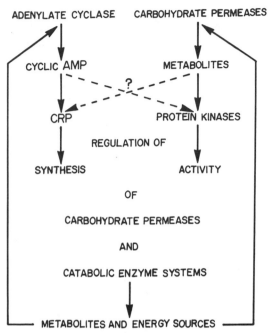

Fig. 6.5. Scheme illustrating what is known regarding the involvement of cyclic AMP and protein kinases in the regulation of carbohydrate catabolism in bacteria. Cyclic AMP acts through the cyclic AMP receptor protein (CRP) to regulate the transcription of carbohydrate catabolic operons encoding carbohydrate permeases and catabolic enzymes. This mechanism has been demonstrated primarily in gram-negative bacteria such as *E. coli*. Intracellular metabolites allosterically regulate protein kinases which phosphorylate carbohydrate transport proteins and catabolic enzymes, thereby regulating their activities. This mechanism has been demonstrated in enteric bacteria for isocitrate dehydrogenase and in gram-positive bacteria for HPr of the phosphotransferase system. The scheme does not exclude the possibility of a regulatory involvement of other effectors such as Ca^{2+} and cyclic GMP. Nor is it meant to suggest that cellular functions involved in carbon metabolism are exclusively regulated by cyclic nucleotides and protein kinases. Flagellar and fimbrial syntheses are known to be regulated by cyclic AMP and CRP in *S. typhimurium,* and protein kinases have been suggested to regulate sporulation in *Bacillus.*

controls the relative activities of the Krebs cycle and the glyoxylate shunt, while phosphorylation of the latter protein apparently controls the rates of inducer uptake and efflux in the presence of a preferable carbon source. The possible involvement of a protein kinase in regulating the activity of the glycolytic enzyme, phosphofructokinase, in *Bacillus* has also been suggested (Price and Gallant,

1983). It is possible that bacterial protein kinases fulfill a diversity of additional cellular functions. For example, an involvement of protein phosphorylation in the sequence of events leading to sporulation in *Bacillus* has not yet been critically examined. Nevertheless, a primary role of protein kinases in the regulation of carbon and energy metabolism in bacteria is becoming apparent.

Cyclic AMP, together with cyclic GMP and Ca^{2+}, functions in eukaryotic cells as a second messenger to regulate the activities of protein kinases which in turn phosphorylate a variety of enzymes and regulatory proteins, thereby influencing their activities. Such a role for these agents has not yet been demonstrated in bacteria. Instead, while the known bacterial protein kinases are apparently sensitive to allosteric control by intracellular metabolites, cyclic AMP functions in gram-negative bacteria through the cyclic AMP receptor protein (CRP) to regulate the transcriptional initiation and translation of genes encoding carbohydrate permeases and catabolic enzyme system (Fig. 6.5). In fact, no known carbohydrate permease gene in *E. coli* (with the exception of the *crrA* gene, encoding the Enzyme III^{glc} of the PTS) is *not* subject to regulation by the cyclic AMP–CRP complex.

A number of other bacterial cell functions are also regulated by the cyclic AMP–CRP complex. These include pigment production in *Serratia marcescens,* bioluminescence in *Vibrio fischeri,* flagellar synthesis in *E. coli,* and fimbrial synthesis in *S. typhimurium* (Saier, 1979; Saier *et al.,* 1978b). It has been argued that all of these cell functions may have originally developed to play a role, either directly or indirectly, in the acquisition of carbon and energy (Saier, 1979). A role for cyclic nucleotides in cellular morphogenesis and cell cyclic control in *Caulobacter crescentus* has also been suggested (Shapiro, 1976; Shapiro *et al.,* 1981), and the same may be true in *E. coli* since cyclic AMP can be bacteriostatic to this organism. In spite of these alternative roles of cyclic AMP in regulatory capacities other than those directly involved in carbon metabolism, it seems clear that cyclic AMP functions primarily in gram-negative bacterial cells as an indicator or "alarmone" of carbon sufficiency.

Many processes in addition to carbon metabolism in eukaryotic cells are influenced by cyclic AMP and protein phosphorylation, and the same possibility cannot be excluded for bacteria. Nevertheless, in view of the findings reported, it seems reasonable to propose that cyclic AMP and protein kinases evolved initially as agents regulating the flow of energy metabolism and only subsequently assumed more complex roles in differentiation and cell maintenance. While the presently recognized mechanisms by which cyclic AMP influences carbon utilization differ in the eukaryotic and prokaryotic kingdoms, further investigations may reveal cyclic AMP-dependent protein kinases in bacteria as well as cyclic AMP-dependent modulation of transcriptional initiation in eukaryotes. It is clear that a profusion of regulatory mechanisms exists for the control of carbon and energy flow. If these mechanisms evolved prior to the divergence of eukaryotes from prokaryotes, they may prove to be universal throughout the living world.

7

Exogenous Induction of Certain Carbohydrate Permeases in Bacteria

Most carbohydrates which can be utilized by bacteria as a principal growth substrate are initially converted to cytoplasmic products which are unique, or nearly unique, to that sugar. These products are only present in the cytoplasm in high concentrations when the parent sugar is transported across the membrane from the extracellular medium. Consequently, they can serve as cytoplasmic inducers of the operons encoding the catabolic enzyme system specific for that sugar. Well-characterized examples include allolactose (derived from lactose), cytoplasmic maltose (derived from exogenous maltose), and α-glycerophosphate (derived from glycerol).

Other carbon sources when transported across the membrane give rise to cytoplasmic products which are normal cellular metabolites. For example, the phosphoglycerate permease and the tricarboxylate permease in *Salmonella* (Saier, 1979; Saier *et al.*, 1975a) actively accumulate cytoplasmic phosphoenolpyruvate and citrate, respectively, and these compounds are normal intermediates of glycolysis and the Krebs cycle, respectively. If such cytoplasmic substances or metabolic products derived from them induced synthesis of the corresponding transport systems, these permeases would always be synthesized at high levels, even when no exogenous substrate was available. The synthesis of unnecessary proteins as well as leakage of essential nutrients from the cell would result in wastage of cellular energy.

To exemplify this point, the phosphoglycerate permease of *S. typhimurium* transports and is induced by 3-phosphoglycerate, 2-phosphoglycerate, and phosphoenolpyruvate (Saier *et al.*, 1975a). Induction of this system occurs when one of these compounds is present in the growth medium at a concentration of 100 μM (Saier *et al.*, 1975a), despite the fact that the log-phase cells contain these three inducers at cytoplasmic concentrations of about 1 mM (Roberts *et al.*, 1963). Since the transport system remains uninduced until one of the substances is added externally, a mechanism involving induction from without is implied.

The tricarboxylic acid permease system of *S. typhimurium* transports the following Krebs cycle tricarboxylates: citrate, isocitrate, tricarballylate, and *cis*- and *trans*-aconitate (Kay and Cameron, 1978). Induction of this system appears to be restricted to citrate, isocitrate, and tricarballylate (Kay and Cameron, 1978; Saier, 1979). As in the case of the phosphoglycerate transport system, the inducers of the system are normal metabolic intermediates which are present within growing cells. However, even in aconitase-negative mutants, in which citrate is accumulated, induction only occurs upon addition of citrate to the external medium. A requirement for extracellular induction was also observed in isocitrate dehydrogenase mutants of *S. typhimurium* (W. W. Kay, personal communication). Under similar conditions, an aconitase mutant of *B. subtilis* showed gratuitous induction (Willecke and Pardee, 1971), but the experiments performed did not rule out the possibility of citrate efflux from the mutant and consequent exogenous induction.

In order to conserve energy, bacteria have evolved mechanisms

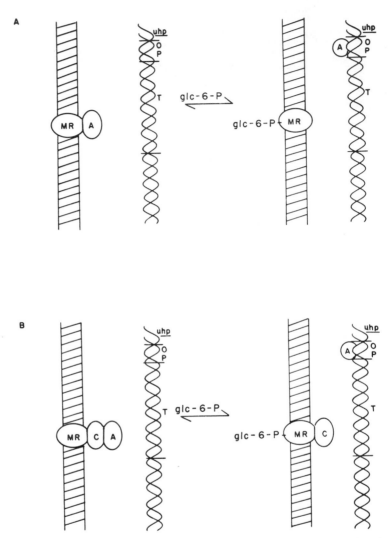

Fig. 7.1 Proposed models for the transcriptional regulation of the hexose phosphate transport (*uhpT*) operon in *E. coli* by extracellular inducer. A. A two-protein regulatory model as originally proposed (Saier, 1979) for the transcriptional regulation of the hexose phosphate or phosphoglycerate transport system in *E. coli* or *S. typhimurium*, respectively. One of the regulatory proteins is an activator protein (A), which possesses affinity for the controller region of the operon encoding the transport protein. It also binds to a transmembrane regulatory protein (MR), which posesses a glucose-6-P bind-

TABLE 7.1

Genes Encoding Transport Proteins Which May Be Expressed
in Response to Exogenous Inducers

Gene designations	Organism	Gene product(s)
uhp	E. coli	Hexose phosphate transport system
pgt	S. typhimurium	Phosphoglycerate transport system
pts	S. typhimurium	Enzyme I and HPr of the PTS
gluA, manA	S. typhimurium	Enzyme II^{glc} and Enzyme II^{man} of the PTS
tct	S. typhimurium	Tricarboxylate transport system

allowing extracellular carbohydrates to serve as inducers of the synthesis of certain permeases. Exogenous induction was first demonstrated for the hexose phosphate permease and later for other permease systems (Table 7.1; Dills *et al.*, 1980). Included in this group
are probably several proteins of the phosphotransferase system
(Table 7.1). In each of the cases to be discussed, the presence of
even high concentrations of the inducer molecule in the cytoplasm is
without effect, but addition of low concentrations of the compound
to the growth medium causes immediate induction.

A. EXOGENOUS INDUCTION OF THE HEXOSE
PHOSPHATE AND PHOSPHOGLYCERATE PERMEASES

In 1979, we proposed a mechanism which accounted for exogenous induction (Saier, 1979). In this model, two regulatory proteins
control the induction process as shown in Fig. 7.1,A. One of these
proteins is a repressor or activator protein (A) which interacts di-

ing site on the external surface of the cytoplasmic membrane. The binding of glucose-
6-P to the external face of the MR protein decreases its affinity for the A protein, allowing
the latter protein to bind to the operator of the *uhpT* operon and activate transcription.
(Modified from Saier, 1979). B. Schematic depiction of a similar model involving an
additional "linker" protein (C) which mediates the binding of A to the MR protein. This
model was recently proposed by Shattuck-Eidens and Kadner (1983).

rectly with the operator region of the target operon. The other is a transmembrane regulatory protein (MR) with a sugar phosphate binding site on the external face of the membrane and a binding site for the A protein on the cytoplasmic side of the membrane. According to this hypothesis, binding of the inducing glucose-6-P molecule to the external domain of the MR protein either enhances affinity of the protein for the A protein (if it is a repressor) or decreases affinity of MR for the A protein (if it is a transcriptional activator). Binding constants and relative concentrations of the MR protein, A protein, and operator site are assumed to be such that exogenous inducer either drains the repressor off the DNA or allows sufficient quantities of the activator to be released from the MR protein so that it can bind to the operator region of the operon encoding the permease. Consequently, transcription of the operon is induced. Other mechanisms of exogenous induction have also been proposed (Dietz, 1976), and an alternative mechanism of exogenous induction has been suggested for the induced synthesis of certain proteins of the phosphotransferase system (Section B). In some cases, transport proteins appear to play a direct role in transcriptional regulation (Saier, 1979; Rephaeli and Saier, 1980b; Saier and Leonard, 1983; see Section C).

The mechanism proposed in the preceding paragraph suggests the existence of at least two regulatory proteins controlling operon expression. One of these regulatory proteins exerts a positive effect on transcription when present in the cell in the native state while the other exerts a negative effect. Regulatory mutants lacking the function of either protein would be either noninducible (absence of the positive effector) or constitutive (absence of the negative effector) for permease synthesis, and one of these two classes of mutants should be epistatic over the other.

Recently, the structural and regulatory genes involved in the synthesis of two exogenously induced permeases have been cloned and studied. One of these is the phosphoglycerate permease (*pgt*) which transports 2- and 3-phosphoglycerate, as well as phosphoenolpyruvate (Saier *et al.*, 1975a). Analyses of the cloned genes have revealed that the structural gene for the permease encodes a 66,000-dalton protein (Hugenholtz *et al.*, 1981). More recently two regulatory genes of the *pgt* system have been identified (J. -S. Hong, personal communication). One appears to encode a positive regulatory element, while the other probably encodes a negative regulatory element as predicted by the proposed model of extracellular induction (Figure 7.1,A).

Comparable studies with the *uhp* system have been reported which also suggest the presence of at least two regulatory genes (Kadner and Shattuck-Eidens, 1983; Shattuck-Eidens and Kadner, 1983). In the former paper over 200 mutants in the *uhp* region were obtained. The defects in these mutants, all of which gave rise to a hexose phosphate-negative phenotype, included point, deletion, and transposon Tn*10* insertion mutations. Additional deletion mutations resulted from imprecise excision of Tn*10* insertions located on either side of the *uhp* region. Conjugal crosses were performed in order to allow determination of the relative order of the *uhp* alleles and the deletion endpoints. In addition, specialized transducing phages carrying the permease structural genes fused to the lactose (*lac*) operon and carrying various amounts of the adjacent *uhp* genes were isolated. These phage were used as genetic donors in crosses which corroborated the results of the conjugal crosses. The proposed gene order was *uhpTRA*, where *uphT* encodes the permease protein and *uhpR* and *uhpA* are proposed regulatory genes whose products are necessary for proper *uhpT* transcription.

Mutations in the *uhpT* or *uhpA* gene reverted with low frequency ($<10^{-7}$ per cell per generation), giving rise only to strains with a normal inducible phenotype. By contrast, *uhpR* mutants reverted with much higher frequencies (sometimes greater than 10^{-5} per cell per generation) and gave rise to constitutive and semiconstitutive expression of *uhpT*. Most of these pseudorevertants retained the original mutations, for example, the original *uhpR*::Tn*10* insertions which conferred the tetracycline resistance of the transposon. This fact and other studies showed that restoration of growth on glucose-6-P was due to secondary mutations within the *uhp* region, at sites distal to the original mutation, possibly within another gene. It was suggested that a second regulatory gene (between *uhpR* and *uhpA*), when defective, gave rise to constitutive expression of *uhpT*.

In a second paper (Shattuck-Eidens and Kadner, 1983), studies in which the *uhp* region was cloned provided further evidence for a positive activator of the expression of *uhpT*. One plasmid (pLC17-47) and its derivatives conferred constitutive glucose-6-P uptake activity to all strains harboring the plasmid, even when the chromosomal *uhp* region was deleted. This same plasmid (and its derivatives) also rendered constitutive the expression of β-galactosidase when encoded by either a chromosomal or plasmid *uhpT–lacZ* gene fusion regardless of the state of the *uhpR* and *uhpA* genes on the chromosome. This plasmid was shown to carry the complete *uhp* region. By contrast, another plasmid, pLC40-33, and its derivatives conferred inducible *uhp* expression which required the presence of the *uhpA*⁺ gene on the chromosome. The induced transport levels in all strains bearing one of these plasmids were not appreciably amplified over the haploid level, and the same behavior was observed with the cloned *uhpT–lacZ* operon fusions. This plasmid appeared to lack an intact *uhpA* gene. In other words, amplification of *uhpA* apparently results in constitutive expression of *uhpT*, regardless of the number of copies of *uhpT* present in the cell, but the presence of a normal copy number of wild-type *uhpA* results in inducible expression of *uhpT*, regardless of the number of copies of either *uhpT* or *uhpR* present within the cell. It was concluded that the *uhpA* gene product is necessary for *uhpT* transcription and that elevated dosages of *uhpA* result in constitutive, or partially constitutive expression of *uhpT*.

To account for the results reported, the following model was proposed: There are three *uhp* regulatory genes in addition to the permease structural gene, *uhpT*. These genes are *uhpR*, *uhpA*, and *uhpC*. The *uhpA* gene product is a positive activator of *uhpT* transcription. In the uninduced state, the *uhpA* gene product may be complexed with the *uhpC* gene regulatory product and therefore be unavailable for transcriptional activation. The third regulatory component, the product of the *uhpR* gene, is the transmembrane protein which binds glucose-6-P on the external surface. Binding of external glucose-6-P alters the conformation of the internal domain which triggers the release of the *uhpA* activator from its inactive complex by interaction with the *uhpC* gene product. This model is shown in Figure 7.1,B.

If this model is correct, then mutational loss of *uhpA* or *uhpR* gives rise to a *uhp*⁻ phenotype. While reversion of *uhpA* should be a rare event, resulting only from restoration of *uhpA* function or an alteration in the *uhpT* promoter, reversion of *uhpR* should be frequent, resulting from mutations in *uhpC* which destroy the function of the *uhpC* protein. Such *uhpC*⁻ mutants should be constitutive regardless of the state of *uhpR*. On the other hand, while *uhpC* is epistatic over *uhpR*, *uhpA* would be expected to be epistatic over both *uhpR* and *uhpC*. It is worth noting that this model does not easily explain the constitutivity of *uhpA* overproducing strains, since plasmids bearing the complete *uhp* region would be expected to overproduce both the *uhpA* and the *uhpC* gene products, and the latter should neutralize the former. It is possible, however, that overexpression of both genes results in the presence of sufficient uncomplexed *uhpA* gene product to allow high constitutive (or partially constitutive) expression of *uhpT*.

At the present time, the only bacteria which have been shown to exhibit exogenous induction mechanisms are the enteric bacteria, *E. coli* and *S. typhimurium*. It seems likely, however, that other bacteria will be shown to exhibit the phenomenon. For example, Hunt and Phibbs (1983) have recently shown that *Pseudomonas aeruginosa* can utilize exogenous sources of glucose, gluconate, and 2-ketogluconate. The pathways proposed for the catabolism of these carbohydrates are shown in Fig. 7.2. Since glucose dehydrogenase and gluconate dehydrogenase act on extracellular substrates and

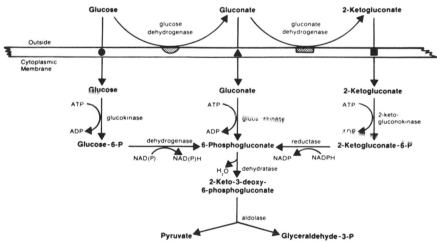

Fig. 7.2. Alternative pathways for the catabolism of exogenous glucose, gluconate, and 2-ketogluconate in *Pseudomonas aeruginosa*. Glucose can either be taken up by the cells or oxidized on the external surface of the membrane by glucose dehydrogenase. Similarly, gluconate can either be taken up or further oxidized to 2-ketogluconate on the outside of the cell. Exogenous 2-ketogluconate can be transported across the membrane and metabolized intracellularly. Synthesis of the gluconate and 2-ketogluconate transport systems as well as glucokinase and gluconate dehydrogenase may be regulated by an exogenous induction mechanism such as that proposed in Fig. 7.1 for the *uhp* operon in *E. coli*. Taken from Hunt and Phibbs 1981, with permission.

release their products into the extracellular medium, it seems reasonable to suggest that these enzymes as well as the gluconate and 2-ketogluconate permeases are induced by exogenous, rather than endogenous inducers. Induction by an endogenous mechanism would cause induction of these enzymes even when an extracellular substrate is not available. Further experiments will be required to ascertain how widespread exogenous induction mechanisms are in the prokaryotic world.

Finally, it should be noted that induction of transcription by external stimuli (cell–cell contacts; hormone receptor occupancy) is well documented in animal cells (Saier and Jacobson, 1984). T-cell-independent immunoglobulin induction (Abramoff and LaVie, 1970) may represent another example of cell surface-mediated transcriptional regulation. It may be that exogenous induction mechanisms are widespread throughout the living world. The possibility of similar mechanisms controlling carbohydrate utilization in bacteria and the induction of tissue-specific proteins in animal cells is intriguing.

B. AUTOGENOUS INDUCTION OF THE PROTEINS OF THE PHOSPHOTRANSFERASE SYSTEM

Among the transport proteins which appear to be induced by exogenous substrates are several of the proteins of the phosphotransferase system. Evidence suggests that expression of the structural genes of the *pts* operon, which encode HPr and Enzyme I, as well as the *gluA* (*ptsG*) and *manA* (*ptsM*) genes, which encode the glucose and mannose Enzymes II, respectively, are induced exogenously (Table 7.1).

Expression of the *pts* operon and the genes coding for these Enzymes II were shown to be induced maximally by glucose. However, when uninduced cells were grown on one of the disaccharides, maltose or melibiose, which releases free glucose inside the cell, no induction occurred. This result indicates that intracellular glucose is not responsible for induction of the phosphotransferase proteins (Rephaeli and Saier, 1980b).

Careful examination of the induction characteristics of the *pts*, *gluA*, and *manA* genes revealed the following (Rephaeli and Saier, 1980b): (1) Induction was dependent on the extracellular sugar substrate of the PTS, as well as the Enzyme II specific for the inducing sugar. Genetic loss of the Enzyme II resulted in a noninducible phenotype with respect to that particular sugar substrate. (2) Induced synthesis of all four proteins was dependent on cyclic AMP and a functional cyclic AMP receptor protein. In the absence of either agent, the rate of protein synthesis was basal or subbasal. (3) Several extracellular substrates of the PTS induced expression of the *pts* operon; fewer sugars induced expression of the *gluA* and *manA* genes. However, in all cases the best inducers appeared to be the most rapidly transported substrates of the PTS, and poor substrates were poor inducers. (4) Loss of either Enzyme I or HPr activity as a result of point mutations in the structural genes for these proteins gave rise to high-level constitutive synthesis of the remaining three proteins under consideration. Deletion of the *pts* and *crr* genes also resulted in constitutive synthesis of Enzyme II^{glc} and Enzyme II^{man}. It is interesting to note that *pts* mutants of *Bacillus subtilis* apparently also synthesize the glucose Enzyme II at a high constitutive level (P. Gay, personal communication). These results have led to the postulation of a mechanism which involves phosphorylation of a regulatory protein that can interact with the deoxyribonucleic acid to regulate transcription (Fig. 7.3). As proposed, phosphorylated RPr is a transcriptional repressor [negative control (Fig. 7.3, left)] and/or free RPr is an activator [positive control (Fig. 7.3, right)] A third possibility, that phosphorylated HPr acts as a repressor, has not been eliminated. Further experiments must be performed to determine whether the hypothetical model proposed in Fig. 7.3 truly accounts for the inducible expression of the *pts* operon.

In addition to the *pts*, *glu*, and *man* operons, some evidence has been presented which was interpreted to suggest that *mtl* operon expression may be autogenously regulated by a mechanism involving the mannitol Enzyme II (Saier and Leonard, 1983). This evidence resulted from a study in which the different functions of the Enzyme II^{mtl} were genetically dissected (Leonard and Saier, 1981). Sixty *mtlA* mutants were characterized with respect to mannitol transport, phosphorylation, and chemotaxis. The majority of the mutants either lacked all the activities attributed to the Enzyme II^{mtl}, or they showed depressed levels of these activities. These mutants were found to be either noninducible (Class I) or poorly inducible for the synthesis of the mannitol-1-P dehydrogenase activity. Among Class I mutants were those which presumably lacked the intact En-

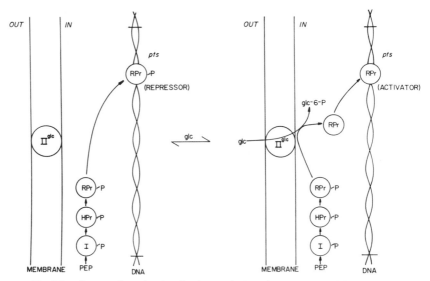

Fig. 7.3. Proposed mechanism for the regulation of transcription of the *pts* operon in *S. typhimurium*. The model suggests that a regulatory protein (here designated RPr) which can be phosphorylated at the expense of phospho-HPr and can interact directly with the operator region of the *pts* operon. Phospho-RPr may be a repressor protein, and/or free RPr may be an activator. The addition of a substrate of the PTS to the cell suspension would be expected to drain phosphate off the phosphorylated, energy-coupling proteins of the PTS with concomitant phosphorylation of sugar. Hence, phospho-RPr would be dephosphorylated to RPr, and transcription of the *pts* operon would be induced. The data presently available do not rule out other possibilities. Abbreviations: glc, glucose; II^{glc}, Enzyme II specific for glucose and methyl-α-glucoside; PEP, phosphoenolpyruvate; glc-6-P, glucose 6-phosphate; I, Enzyme I; DNA, deoxyribonucleic acid. [From Dills *et al.* (1980), with permission.]

zyme II^{mtl} protein because of nonsense mutations in the *mtlA* structural gene. One class of mutants (Class II) lost all activites of the Enzyme II^{mtl} except for the mannitol:mannitol-1-P transphosphorylation activity (see Leonard and Saier, 1981 and Chapter 3, Section E). Cultures of these mutants were examined in the late logarithmic phase or stationary phase of growth and were found to synthesize the mannitol-1-P dehydrogenase at high constitutive levels. This observation led to the suggestion that the Enzyme II^{mtl} somehow plays a role in the transcriptional regulation of the *mtl* operon by an autogenous mechanism (Saier and Leonard, 1983).

A second observation was that, as for the *pts, glu,* and *man* operons, stationary phase cultures of all "tight" *pts* mutants expressed the *mtl* operon constitutively. Two alternative interpretations were considered. First, it was possible that the signal for dephosphorylation of a PTS protein was the signal for induction of *mtl* operon expression. If this were the case, then the constitutivity of class II *mtlA* mutants pointed to the Enzyme II^{mtl} as the PTS protein in question. It should be noted, however, that transrecessive, temperature-sensitive mutations leading to constitutive expression of the mannitol operon have been isolated (Lengeler and Steinberger, 1978). The defective gene has been designated *mtlR*. Any model for the transcriptional regulation of the *mtl* operon must take these mutants into account.

The alternative interpretation suggests that *pts* or Class II *mtlA* mutations cause inducer [presumed to be free mannitol, based on experiments described by Lengeler and Steinberger (1978)] to accumulate in the cell cytoplasm. This inducer would have to arise by reduction of fructose-6-P to mannitol-1-P in a process coupled to NADH oxidation and catalyzed by mannitol-1-P dehydrogenase, followed by hydrolysis of mannitol-1-P to free mannitol. This second process might conceivably be catalyzed by the Enzyme II^{mtl}, thus accounting for the *mtl* operon constitutivity of Class II *mtlA* mutants. *E. coli* cells growing on a carbon source other than mannitol have been shown to accumulate mannitol-1-P under certain physiological conditions (H. Rosenberg, personal communication). If *pts* mutations were to enhance the level of cytoplasmic mannitol-1-P, the generation of endogenous inducer could account for the anomalous regulatory characteristics of the *pts* and Class II *mtlA* mutants.

Experiments have recently been performed which argue against an autogenous induction mechanism involving the Enzyme II^{mtl}. These experiments strongly suggest that the accumulation of free biosynthetic mannitol, during growth on other carbon sources, explains the constitutive expression of the *mtl* genes in *S. typhimurium* (M. J. Novotny and M. H. Saier, unpublished results). In these experiments, wild-type bacteria, Class II *mtlA* mutants, and strains deleted for the *pts* operon were grown in a mineral salts medium containing lactate and succinate as carbon source. Cells were grown in a large volume from a small inoculum. When the bacteria were harvested after extended growth at low cell density, conditions which should minimize accumulation of biosynthetic mannitol, low basal activity of the mannitol-1-P dehydrogenase was observed.

Only when the cultures approached high cell densities were enhanced levels of the dehydrogenase observed. These observations are consistent with the conclusion of Lengeler and Steinberger (1978) that free cytoplasmic mannitol serves as the inducer of the *mtl* operon by a conventional repressor-type mechanism. Defective PTS function presumably prevents phosphorylation of biosynthetic mannitol to mannitol-1-P, thus accounting for the lack of induction of these mutants. The apparent constitutive expression of the *mtl* operon in Class II mutants may have resulted because only the latter mutants have retained the mannitol-1-P recognition site and therefore are capable of hydrolyzing mannitol-1-P to cytoplasmic inducer.

C. POSSIBLE INVOLVEMENT OF TRANSPORT PROTEINS IN TRANSCRIPTIONAL REGULATION

In addition to the PTS proteins discussed in the previous section, several protein constituents of transport systems have been implicated in transcriptional regulation. These proteins include (1) the lactose Enzyme III of the *Staphylococcus aureus* PTS, (2) the *malK* constitutent of the maltose permease in *E. coli*, and (3) the Enzyme III^{glc} of the PTS in enteric bacteria. In none of these cases are the regulatory mechanisms known, and the evidence for their involvement in transcriptional regulation is only indirect. Nevertheless, because of the conceptual importance of studies which define the multifunctional natures of these proteins, this work will be briefly presented.

Lactose is transported across the cytoplasmic membrane of *Staphylococcus aureus* by the lactose phosphotransferase system (see Chapter 6). The uptake and phosphorylation of lactose via this system requires the functional integrity of Enzyme I, HPr, Enzyme II^{lac}, and Enzyme III^{lac}. The product, cytoplasmic lactose 6-phosphate, is then cleaved to glucose and galactose 6-phosphate by a phospho-β-galactosidase. The latter compound is apparently the true inducer of the regulon which codes for the lactose-specific constitutents of the system, Enzyme II^{lac}, Enzyme III^{lac}, and the phospho-β-galactosidase. This phosphate ester can induce the operon when added to a cell suspension at high concentration, even when the phosphotransferase system is nonfunctional.

Mutants defective for lactose utilization were isolated from two *S. aureus* parental strains (Simoni and Roseman, 1973). One was the lactose inducible wild-type strain, and the other was a lactose constitutive strain derived from the inducible parent. The two strains each yielded three identical classes of lactose-negative mutants defective in either Enzyme I, Enzyme II^{lac}, or phospho-β-galactosidase. When the inducible parent was studied, a fourth class of mutants was found that lacked all three of the proteins of the lactose regulon, Enzyme II^{lac}, Enzyme III^{lac}, and phospho-β-galactosidase. Extensive attempts to isolate mutants lacking only Enzyme III^{lac} from the inducible parent gave negative results. By contrast, the constitutive parental strain did not yield mutants lacking the three proteins of the lactose regulon but yielded a major class lacking only Enzyme III^{lac}. The difference in behavior of the inducible and constitutive strains suggested that Enzyme III^{lac} may play a role in regulating transcription or translation (or both) of the lactose regulon in *S. aureus* in addition to its role in the transport of lactose.

Extensive genetic studies on maltose utilization in *E. coli* provided evidence for a second example in which a transport protein possibly functions directly in the regulation of gene expression. The maltose catabolic enzyme system is coded for by genes which are included within four operons (see Chapter 2, Section D). These operons comprise the maltose regulon and are subject to control by the product of the *malT* gene. In the absence of the *malT* protein, expression of the maltose regulon is suppressed. The structural genes, *malP* and *malQ* in the *malA* region (at 74 minutes on the *E. coli* chromosome) code for maltodextrin phosphorylase and amylomaltase, respectively (Hatfield *et al.*, 1969; Bedouelle, 1984). The structural genes *malE*, *malF*, *malG*, and *malK* encode four essential constituents of the maltose permease.

Mutations in these genes have startling effects on the expression of the *malPQ* operon. Loss of *malE*, *malF*, or *malG* function renders the catabolic enzymes noninducible or poorly inducible, but mutation of the *malK* gene results in a level of constitutive expression of the *malPQ* operon equal to about 50% of the maximally induced wild-type level (Hatfield *et al.*, 1969). Although other interpretations can be entertained, this unexpected result suggests that the protein product of the *malK* gene may play a direct role in regulating the expression of the maltose regulon.

Finally, mutations in the *crrA* gene which encodes the Enzyme

III^{glc} (see Chapter 4, Section C) have pleiotropic effects on the expression of numerous operons in *Salmonella typhimurium* (Saier and Roseman, 1976a). One consequence of these mutations is to depress the growth rates on a number of compounds including acetate, α-ketoglutarate, succinate, L-alanine, and guanosine. Secondary mutations within a closely linked gene, designated *crrB*, pleiotropically overcomes this growth defect (A. M. Chin, R. C. Tuttle, J. Loken, and M. H. Saier, Jr., unpublished results). It has been shown that while *pts crrA* double mutants do not induce synthesis of the transport systems and catabolic enzymes responsible for the utilization of the above mentioned carbon sources, induction occurs normally in the *pts crrA crrB* triple mutants. Among the enzymes examined were L-alanine dehydrogenase and succinate dehydrogenase. While the nature of the transcriptional regulatory mechanisms influenced by the *crrB* gene product is not known, the results implicate the PTS in a novel pleiotropic mechanism for the regulation of carbon utilization. Further studies will be required to establish the mechanistic details.

Permease Classification and Mechanisms: Conclusions and Future Perspectives

A. DIVERSITY OF PERMEASE MECHANISMS IN LIVING CELLS

While the bacterial carbohydrate permeases discussed in this volume are clearly among the most thoroughly investigated transport systems, other systems, particularly some of those in animal cells, have been subjected to critical scrutiny. It is instructive to classify transmembrane solute transport systems found in all living cells and to compare their properties with the bacterial systems examined in this volume. In Table 8.1, these permeation systems are grouped into seven major classes, and in Table 8.2, the characteristics of these systems are summarized.

1. Nonstereospecific channel proteins include the ion channel complexes of nerve and muscle cells, the porin protein complexes in the outer membranes of gram-negative bacteria, and the glycerol permease in the *E. coli* inner membrane. These transport proteins

TABLE 8.1

Well-Characterized Biological Solute Transport Systems[a]

I. Nonstereospecific channel proteins
A. Acetylcholine receptor
B. Nerve Na^+ channel
C. Nerve K^+ channel
D. Porins in the outer membranes of gram-negative bacteria
E. Glycerol permease in *E. coli*
II. Single species facilitators (unidirectional or antiport)
A. Glucose transporter in red cell
B. Anion antiporter in red cell
C. Na^+/H^+ antiporter in bacteria and tissue cells
D. ADP/ATP antiporter in mitochondria
III. Two species facilitators (symporters; chemiosmotically coupled)
A. Lactose permease in *E. coli* (lactose : H^+)
B. Melibiose permease in *Salmonella* (melibiose : Na^+)
C. Glucose transporter in certain epithelial cells (glucose : Na^+)
D. A-amino acid transporter in tissue cells (amino acid : Na^+)
E. NaCl/KCl symporter in animal cells
IV. ATPase-coupled active transporters
A. Na^+, K^+-ATPase in animal cells
B. K^+-ATPase in *E. coli*
C. Ca^{2+}-ATPase in animal cells
D. H^+-ATPase in bacteria and mitochondria
V. Periplasmic binding protein–dependent systems
A. Maltose permease in *E. coli*
B. Histidine permease in *Salmonella*
C. Succinate permease in *E. coli*
VI. Light and electron flow driven H^+ or Cl^- transporters
A. Bacteriorhodopsin
B. Halorhodopsin
C. Cytochrome oxidase
VII. Group translocators
A. Mannitol Enzyme II of *E. coli* phosphotransferase system
B. Glucose Enzyme II of *E. coli* phosphotransferase system

[a] For references concerning each of these transport systems, see Saier and Boyden (1984).

TABLE 8.2

Characteristics of Different Classes of Transport Systems

Transport classification (from Table 8.1)	Transport characteristic						
	Solute co-transport	High degree of solute specificity	Readily reversible	(Accelerative) exchange transport	Exchange transport in absence of a substrate	Energy utilization stoichiometric	Single polypeptide chain catalyzes transport
I	−	− or+	+	−	−	−	− or +
II	−	+	+	+	N.R.[a]	−	+
III	+	+	+	+	−	−	+
IV	−	+	−	−	+	+	−
V	−	+	−	−	−	−	−
VI	−	+	−	−	−	+	+ or−
VII (unidirectional)	−	+	−	−	−	+	+
VII (bidirectional)	−	+	+	+	N.R.[a]	−	+

[a] N.R. = Not relevant.

never couple transport of one solute to that of another or catalyze accelerative exchange transport. Their activities follow diffusion kinetics.

2. Single-species facilitators can only catalyze translocation of a single molecular species in one direction at any one time. They may catalyze either unidirectional transport or bidirectional accelerative exchange transport as in the case of the glucose transport system in the human erythrocyte, or they may function as obligatory antiport systems, as in the case of the anion exchanger in the red cell and the Na^+/H^+ antiporter in bacteria and animal cells. Carbohydrates are not transported by this mechanism in *E. coli,* but other bacteria may utilize this simple process.

3. Two-species facilitators (symporters) resemble single-species facilitators in all respects except for their co-transport characteristic. The best characterized of these systems is the lactose permease in *E. coli.* Each of these systems may consist of a single polypeptide chain and function by a carrier-type mechanism. Normally, solute–cation transport is obligatorily coupled, but mutants have been isolated which convert the lactose:H^+ symporter (Class III) into a lactose uniporter (Class II). The characteristics of these mutants emphasize the close mechanistic relationship between class II and class III permeases.

4. Chemically coupled active transporters include cation translocating ATPases. With the possible exception of the Ca^{2+}-ATPase, each of these permease complexes consist of more than one type of polypeptide chain. These transport systems differ from facilitators in catalyzing essentially irreversible translocation processes in the presence of the appropriate energy source. Accelerative exchange transport is not observed. However, the ion translocating ATPases can catalyze single species exchange in the absence of the other ion(s) transported.

5. Periplasmic binding protein-dependent systems of gram-negative bacteria appear to depend on the presence of three or four distinct proteins. The water soluble solute binding protein constituent is in the periplasmic space, two proteins span the membrane, and the fourth protein may be bound to the permease complex on the cytoplasmic surface of the membrane. A pore-type mechanism involving two distinct substrate binding sites, one on each of the two sides of the membrane and each associated with one of the integral proteins of the system, has been proposed for the succinate per-

mease. A fairly nonspecific channel has been proposed for the histidine permease.

6. Some active translocators catalyze unidirectional ion transport at the expense of nonchemical energy sources. Bacteriorhodopsin and halorhodopsin, which use solar energy to drive proton and chloride transport, respectively, and cytochrome oxidase, which obtains energy for proton extrusion from electron transport, are well-characterized examples. Proton translocation via sequential "jumps" to polar sites within bacteriorhodopsin may effect proton transport in this protein. The equivalence of a channel-type mechanism is probably operative.

7. Finally, the sugar-specific components of the bacterial phosphotransferase system, the integral membrane Enzymes II of this complex enzyme system, catalyze two distinct processes. Unidirectional sugar translocation is coupled to the phosphoenolpyruvate-dependent phosphorylation of sugar in a process requiring the energy coupling proteins of the system, while bidirectional transport, catalyzed exclusively by the Enzyme II, is analogous to accelerative exchange transport except that it is coupled to a "transphosphorylation" reaction. In this process, the phosphoryl group of an intracellular sugar phosphate is transferred to the incoming sugar while the sugar moiety of the intracellular sugar phosphate is expelled from the cell. Bidirectional group translocation superficially resembles the process of exchange transport catalyzed by antiporters (Class II transport systems in Table 8.1).

While considerable uncertainty exists with respect to the mechanism by which each of these permeases translocates its substrate across the membrane, it is clear that several fundamentally distinct transport mechanisms exist. Almost every class of transport system is represented by those found in *E. coli*. As *E. coli* is the organism for which genetic, physiological, and biochemical manipulations are most easily performed and highly refined, *E. coli* is clearly the organism of choice for transmembrane solute transport studies.

B. CLASSIFICATION OF CARBOHYDRATE TRANSPORT SYSTEMS IN *E. COLI* AND *S. TYPHIMURIUM*

Table 8.2 summarizes the functional characteristics of the various permease classes, while Table 8.3 groups the known carbohydrate

TABLE 8.3

Classification of Known Carbohydrate Permeases in *E. coli* and *S. typhimurium*[a,b]

Class I	Class III	Class V	Class VII
Glycerol	L-Arabinose	L-Arabinose	N-Acetylglucosamine
Outer	Galactose	Dicarboxylates	Dihydroxyacetone
membrane	Gluconate	Maltose	Fructose
porins	Glucuronate	Methyl β-galactoside	Galactitol
	α-Glycerophosphate	Ribose	Glucitol
	Hexose phosphate	Tricarboxylates[d]	Glucose
	Lactate	Xylose	β-Glucosides[c]
	Lactose[c]		Mannitol
	Melibiose		Mannose
	Phosphoglycerate[d]		Sucrose[e]
	Pyruvate		Trehalose
	Xylose		

[a] The periplasmic sugar-binding protein components of class V permeases and the Enzyme II components of Class VII permeases apparently all serve as chemoreceptors. No component of the Class I or Class III sugar permeases serves as a chemoreceptor (Adler *et al.*, 1973).

[b] Carbohydrates utilized by *S. typhimurium* as sole carbon source for which the relevant transport system has not been adequately characterized include galactarate, galacturonate, glucarate, glycerate, glycolate, inositol, α-ketoglutarate, L-lyxose and rhamnose (Gutnick *et al.*, 1969). The table does not include fatty acids, nucleosides and amino acids.

[c] These systems are found only in *E. coli*.

[d] These systems are found only in *S. typhimurium*.

[e] This system is plasmid encoded in some *Salmonella* strains.

permeases in *E. coli* and *S. typhimurium* according to their characteristics in order to classify each of these systems within one of the seven categories defined in the preceding section. In these two organisms, one nonspecific pore protein in the cytoplasmic membrane and several porin proteins in the outer membrane (see Table 1.1) comprise the Class I permeases. Twelve carbohydrate:cation symport systems make up the group of Class III permeases. Seven periplasmic solute-binding protein-dependent systems (Class V) are known, and eleven PTS-dependent sugar permeases (Enzymes II; Class VII), including one plasmid-encoded system, have been characterized. Many similarities will undoubtedly emerge as the structures and mechanistic features of the permeases within each group become known. A most interesting aspect of future comparative studies will be to define the structural and mechanistic differences within each group as well as the similarities among permeases of different groups. It is also possible that permeases not yet characterized (see Footnote b in Table 8.3) will prove to function by mechanisms which do not fall into any one of these four categories.

Finally, it is worth noting that components of each of the Class V and Class VII permeases (the periplasmic solute-binding proteins and Enzymes II, respectively) assume the additional function of chemoreception although Class I and III proteins do not. The two chemotactic mechanisms represented by the chemotactic Class V and VII permeases appear to be quite different. Adaptation to Class V substrates involves methylation while adaptation to Class VII substrates does not (Niwano and Taylor, 1982). The interactions of these proteins with elements of the sensory transmission systems which direct flagellar rotation are essentially undefined. A more primitive mechanism of chemoreception would be predicted for the proteins of the phosphotransferase system if this system evolved prior to the appearance of oxygen on earth (Saier, 1977). Possibly sugar transport, chemoreception, and phosphorylation mediated by the PTS, as well as the process by which the system regulates other cellular physiological processes are all interrelated. Such an interrelationship would explain the association of all of these functions with a single protein (see Chapter 3, Section E and Saier and Jacobson, 1984).

C. CONCLUSIONS

During the past 4 years, tremendous progress has been made in defining the structures and physicochemical properties of representative permeases within each group of the major types of bacterial carbohydrate transport systems. For example, the lactose permease and the mannitol Enzyme II of the PTS have been purified to homogeneity and reconstituted in artificial membranes, and the nucleotide sequences of their genes have been determined. From the gene sequences, the amino acid sequences of the proteins have been deduced, and hydropathy analyses of these sequences have yielded interesting predictions regarding the topographies of these proteins in the membrane. Genes encoding most of the maltose permease constituents have also been sequenced leading to some important predictions regarding the cooperative interactions which must ultimately explain the functioning of this permease system. Finally, corresponding structural information for the simplest of the carbohydrate permeases in the *E. coli* cytoplasmic membrane, the glycerol facilitator, should be forthcoming as soon as this gene is cloned.

The notion that both channels (Fig. 1.1, top) and carriers (Fig. 1.1, bottom) account for transport phenomena in biological membranes seems well established. The best-characterized examples of channel-type systems include the outer membrane porins discussed in Chapter 1 and the glycerol pore discussed in Chapter 2. The best characterized carrier-type permease is probably the lactose permease in *E. coli* (Chapter 2). Yet hybrids of these mechanisms may exist. The fact that the Enzymes II of the PTS catalyze both unidirectional uptake of sugar, coupled with sugar phosphorylation, and unidirectional efflux of free sugar in a much more rapid nonphosphorylative process, lends credence to this possibility. The molecular details of these processes and their interrelationships should prove interesting.

With so much now known about the structures of bacterial permeases, it will be the principal goal of the membrane biochemist to integrate the earlier mechanistic studies with the more recently established structural information. The nature of permease subunit interactions and their possible significance to unidirectional and/or bidirectional transport will undoubtedly become known in the next few years. The possibility, now somewhat out of vogue, that a half-of-sites mechanism might account for transmembrane solute transport must be thoroughly re-evaluated. The nature and identity of the active site aminoacyl residues must be established, and a three-dimensional model must be constructed. The number of solute binding sites and the manner in which they function in translocation of the substrate across the membrane must also be established for each permease system under study. The possibility of stereospecific channels through permease proteins must be rigorously evaluated. Thus, while different types of systems may overlap in their "choice" of translocation mechanism, this can never be assumed *a priori* by the investigator. Both biochemical and genetic approaches will undoubtedly prove useful for the acquisition of this knowledge, and all facts must ultimately be integrated into a coherent whole.

Progress within the past 4 years in understanding the mechanisms regulating the activities of bacterial carbohydrate permeases has been equally impressive. For example, the essential features of the

PTS-mediated regulation of the lactose permease in *E. coli* have become established. The existence of other novel regulatory processes controlling both the uptake of sugars and the extrusion of cytoplasmic sugar phosphates has also been established. The surprising involvement of metabolite controlled protein kinases is not only of great mechanistic significance, it also provides insight into the evolution of those enzymes as regulators of cellular metabolism. Further, the occurrence of exogenous induction mechanisms in bacteria may extrapolate to mechanisms of cell surface activation in higher eukaryotes. These observations establish the biological unifying principles which state that common mechanisms of metabolism, biosynthesis, and regulation will be found universally throughout the living world. Establishment of the fine details of the processes will provide additional clues as to the relatedness and mechanisms of regulatory processes chosen by evolutionarily divergent organisms.

Evolution, like the tinker, does not produce innovations from scratch. It works on what already exists, transforming a system to give it a new function or combining several systems to produce a more complex one.

F. Jacob
from *The Possible and the Actual*

Bibliography

Abramoff, P., and LaVie, M. (1970). "Biology of the Immune Response." McGraw-Hill, New York.

Adler, J., Hazelbauer, G. L., and Dahl, M. M. (1973). Chemotaxis toward sugars in *Escherichia coli*. *J. Bacteriol.* **115,** 824–847.

Ahmed, S., and Booth, I. R. (1981). The effects of partial and selective reduction in the components of the proton-motive force on lactose uptake in *Escherichia coli*. *Biochem. J.* **200,** 583–589.

Alper, M. D., and Ames, B. N. (1975). Cyclic 3′,5′-adenosine monophosphate phosphodiesterase mutants of *Salmonella typhimurium*. *J. Bacteriol.* **122,** 1081–1090.

Amaral, D., and Kornberg, H. L. (1975). Regulation of fructose uptake by glucose in *Escherichia coli*. *J. Gen. Microbiol.* **90,** 157–168.

Anderson, B., Weigel, N., Kundig, W., and Roseman, S. (1971). Sugar transport. III. Purification and properties of a phosphocarrier protein (HPr) of the phosphoenolpyruvate dependent phosphotransferase system of *E. coli*. *J. Biol. Chem.* **246,** 7023–7033.

Bachmann, B. J. (1983). Linkage map of *Escherichia coli* K-12. Edition 7. *Microbiol. Rev.* **47,** 180–230.

Bavoil, P., and Nikaido, H. (1981). Physical interaction between the phage receptor protein and the carrier-immobilized maltose-binding protein of *Escherichia coli*. *J. Biol. Chem.* **256,** 11385–11388.

Bavoil, P., Hofnung, M., and Nikaido, H. (1980). Identification of a cytoplasmic membrane-associated component of the maltose transport system of *Escherichia coli, J. Biol. Chem.* **255,** 8366–8369.

Bavoil, P., Wandersman, C., Schwartz, M., and Nikaido, H. (1983). A mutant form of maltose-binding protein of *Escherichia coli* deficient in its interaction with the phage lambda receptor protein. *J. Bacteriol.* **155,** 919–921.

Bedouelle, H. (1984). Controle de l'utilisation du maltose et des mactodextrines par *Escherichia coli, Bull. de l'Inst. Pasteur* **82,** 91–145.

Begley, G. S., Hansen, D. E., Jacobson, G. R., and Knowles, J. R. (1982). Stereochemical course of the reactions catalyzed by the bacterial phosphoenolpyruvate : glucose phosphotransferase system. *Biochemistry* **21,** 5552–5556.

Beneski, D. A., Nakazawa, A., Weigel, N., Hartman, P. E., and Roseman, S. (1982). Sugar transport by the bacterial phosphotransferase system. XIV. Isolation and characterization of a phosphocarrier protein HPr from wild type and mutants of *Salamonella typhimurium. J. Biol. Chem.* **257,** 14492–14498.

Berger, E. A., and Heppel, L. A. (1974). Different mechanisms of energy coupling for the shock-sensitive and shock-resistant amino acid permeases of *Escherichia coli. J. Biol. Chem.* **249,** 7747–7755.

Berman-Kurtz, M., Lin, E. C. C., and Richey, D. P. (1971). Promoter-like mutant with increased expression of the glycerol kinase operon of *Escherichia coli. J. Bacteriol.* **106,** 724–731.

Bewick, M. A., and Lo, T. C. Y. (1979). Dicarboxylic acid transport in *Escherichia coli* K12: involvement of a binding protein in the translocation of dicarboxylic acids across the outer membrane of the cell envelope. *Can. J. Biochem.* **57,** 653–661.

Beyreuther, K., Raufuss, H., Schrecken, O., and Hengstenberg, W. (1977). The phosphoenolpyruvate-dependent phosphotransferase system of *Staphylococcus aureus*. 1. Amino acid sequence of the phosphocarrier protein HPr. *Eur. J. Biochem.* **75,** 275–286.

Bitoun, R., de Reuse, H., Touati-Schwartz, D., and Danchin, A. (1983). The phosphoenolpyruvate dependent carbohydrate phosphotransferase system of *Escherichia coli. FEMS Microbiol. Lett.* **16,** 163–167.

Bocklage, J., and Müller-Hill, B. (1983). lacZ⁻–Y⁺ fusions in *Escherichia coli,* DNA sequencing reveals the eight N-terminal residues of *lac* permease as nonessential. *Eur. J. Biochem.* **137,** 561–565.

Boos, W. (1982). Aspects of maltose transport in *Escherichia coli:* Established facts and educated guesses. *Ann. Microbiol. (Paris)* **133A,** 145–152.

Booth, I. R., Kroll, R. G., and Ahmed, S. (1981). Generation and utilization of the proton motive force in *Escherichia coli. Dev. Bioenerg. Biomembr.* **5,** 423–426.

Bourgeois, S., and Pfahl, M. (1976). Repressors. *Adv. Protein Chem.* **30,** 1–99.

Brana, H., and Chytil, F. (1966). Splitting of the cyclic 3′,5′-adenosine monophosphate in cell-free system of *Escherichia coli. Folia Microbiol. (Prague)* **11,** 43–46.

Brass, J. M. (1982). Reconstitution of maltose transport in *malB* and *malA* mutants of *Escherichia coli*. *Ann. Microbiol.* (*Paris*) **133A**, 171–180.

Brass, J. M., Ehmann, U., and Bukau, B. (1983). Reconstitution of maltose transport in *Escherichia coli:* conditions affecting import of maltose-binding protein into the periplasm of calcium-treated cells. *J. Bacteriol.* **155**, 97–106.

Britton, P., Boronat, A., Hartley, D. A., Jones-Mortimer, M. C., Kornberg, H. L., and Parra, F. (1983). Phosphotransferase-mediated regulation of carbohydrate utilization in *Escherichia coli* K12: Location of the *gsr* (*tgs*) and *iex* (*crr*) genes by specialized transduction. *J. Gen. Microbiol.* **129**, 349–358.

Büchel, D. E., Gronenborn, B., and Müller-Hill, B. (1980). Sequence of the lactose permease gene. *Nature* (*London*) **283**, 541–545.

Burd, G. I., Gabrielyan, T. R., Bol'shakova, T. N., and Gershanovich, V. N. (1980). Study of linkage with *pts* genes of pleiotropic mutation affecting expression of catabolite-sensitive genes in *Escherichia coli* K-12. *Sov. Genet.* (*Engl. Transl.*) **16**, 622–629.

Caban, C. E., and Ginsberg, A. (1976). Glutamine synthetase adenylyl transferase from *Escherichia coli:* Purification and physical and chemical properties. *Biochemistry* **15**, 1569–1580.

Carrasco, N., Tahara, S. M., Patel, L., Goldkorn, T., and Kaback, H. R. (1982). Preparation, characterization, and properties of monoclonal antibodies against the *lac* carrier protein from *Escherichia coli*. *Proc. Natl. Acad. Sci. U.S.A.* **79**, 6894–6898.

Casadaban, M. J., and Cohen, S. N. (1979). Lactose genes fused to exogenous promoters in one step using a Mu-*lac* bacteriophage: *In vivo* probe for transcriptional control sequences. *Proc. Natl. Acad. Sci. U.S.A.* **76**, 4530–4533.

Castro, L., Feucht, B. U., Morse, M. L., and Saier, M. H., Jr. (1976). Regulation of carbohydrate permeases and adenylate cyclase in *Escherichia coli*. *J. Biol. Chem.* **251**, 5522–5527.

Chapman, A. G., Fall, L., and Alkinson, D. E. (1971). Adenylate energy charge in *Escherichia coli* during growth and starvation. *J. Bacteriol.* **108**, 1072–1086.

Chapon, C. (1982). Role of the catabolite activator protein in the expression of the maltose regulon of *Escherichia coli*. *Ann. Microbiol.* (*Paris*) **133A**, 77–80.

Chou, P. Y., and Fasman, G. D. (1978). Empirical predictions of protein conformation. *Annu. Rev. Biochem.* **47**, 251–276.

Clement, J. M., Braun-Breton, C., Lepouce, E., Marchal, C., Perrin, D., Villarroya, H., and Hofnung, M. (1982). A system for genetic analysis in gene *lamB:* first results with phage lambda resistant *tight* mutants. *Ann. Microbiol.* (*Paris*) **133A**, 9–20.

Cohn, D. E., and Kaback, H. R. (1980). Mechanism of the melibiose porter in membrane vesicles of *Escherichia coli*. *Biochemistry* **19**, 4237–4243.

Cohn, D. E., Kaczorowski, G. J., and Kaback, H. R. (1981). Effect of the proton electrochemical gradient on maleimide inactivation of active transport in *Escherichia coli* membrane vesicles. *Biochemistry* **20**, 3308–3313.

Conrad, C. A., Sterns, G. W., III, Prater, W. E., Rheiner, J. A., and Johnson, J. R. (1984). Characterization of a *glpK* transducing phage. *Mol. Gen. Genet.* **193**, 376–378.

Cordaro, J. C., and Roseman, S. (1972). Deletion mapping of the genes coding for

HPr and Enzyme I of the phosphoenolpyruvate : sugar phosphotransferase system in *Salmonella typhimurium. J. Bacteriol.* **122,** 17–29.

Cordaro, J. C., Melton, T., Stratis, J. P., Atagun, M., Gladding, C., Hartman, P. E., and Roseman, S. (1976). Fosfomycin resistance: Selection method for internal and extended deletions of the phosphoenolpyruvate: Sugar phosphotransferase genes of *Salmonella typhimurium. J. Bacteriol.* **128,** 785–793.

Cozzarelli, N. R., Freedberg, W. B., and Lin, E. C. C. (1968). Genetic control of the L-α-glycerophosphate system in *Escherichia coli. J. Mol. Biol.* **31,** 371–387.

Dambe, V. N., and Karlin, A. (1978). Affinity labeling of one of two α-neurotoxin binding sites in acetylcholine receptor from *Torpedo californica. Biochemistry* **17,** 2039–2045.

Daniel, J. (1984). Enzyme III stimulation of cyclic AMP synthesis in an *Escherichia coli crp* mutant. *J. Bacteriol.* **157,** 940–941.

Daniel, J., Dondon, L., and Danchin, A. (1983). 2-Ketobutyrate: A putative alarmone of *Escherichia coli. Mol. Gen. Genet.* **190,** 452–458.

Delbaere, L. T. J., Bruse, L. M., and Waygood, E. B. (1982). Preliminary X-ray data for the HPr protein of the phosphoenolpyruvate : sugar phosphotransferase system (PTS) of *Escherichia coli. J. Mol. Biol.* **157,** 161–162.

Demerec, M., Adelberg, E. A., Clark, A. J., and Hartman, P.E. (1966). A proposal for a uniform nomenclature in bacterial genetics. *Genetics* **54,** 61–76.

de Riel, J. K., and Paulus, H. (1978a). Subunit dissociation in the allosteric regulation of glycerol kinase from *Escherichia coli.* I. Kinetic evidence. *Biochemistry* **17,** 5134–5140.

de Riel, J. K., and Paulus, H. (1978b). Subunit dissociation in the allosteric regulation of glycerol kinase from *Escherichia coli.* II. Physical evidence. *Biochemistry* **17,** 5141–5145.

de Riel, J. K., and Paulus, H. (1978c) Subunit dissociation in the allosteric regulation of glycerol kinase from *Escherichia coli.* III. Role in desensitization. *Biochemistry* **17,** 5146–5150.

Deutscher, J., and Saier, M. H., Jr. (1983). ATP-dependent protein kinase-catalyzed phosphorylation of a seryl residue in HPr, the phosphoryl carrier protein of the phosphotransferase system in *Streptococcus pyogenes. Proc. Natl. Acad. Sci. U.S.A.* **80,** 6790–6794.

Deutscher, J., Beyreuther, K., Sobek, H. M., Stüber, K., and Hengstenberg, W. (1982). Phosphoenolpyruvate-dependent phosphotransferase system of *Staphylococcus aureus:* Factor III^{lac}, a trimeric phospho-carrier protein that also acts as a phase transfer catalyst. *Biochemistry* **21,** 4867–4873.

Dietz, G. W., Jr., (1976). The hexose phosphate transport system of *Escherichia coli. Adv. Enzymol.* **44,** 237–259.

Dills, S. S., and Seno, S. (1983). Regulation of hexitol catabolism in *Streptococcus mutans. J. Bacteriol.* **153,** 861–866.

Dills, S. S., Apperson, A., Schmidt, M. R., and Saier, M. H., Jr. (1980). Carbohydrate transport in bacteria. *Microbiol. Rev.* **44,** 385–418.

Dills, S. S., Schmidt, M. R., and Saier, M. H., Jr. (1982). Regulation of lactose transport by the phosphoenolpyruvate–sugar phosphotransferase system in membrane vesicles of *Escherichia coli. J. Cell. Biochem.* **18,** 239–244.

Dobrogosz, W. J., Hall, G. W., Sherba, D. K., Silva, D. O., Harman, J. G., and

Melton, T. (1983). Regulatory interactions among the *cya, crp* and *pts* gene products in *Salmonella typhimurium. Mol. Gen. Genet.* **192**, 477–486.

Dooijewaard, G., Roossien, F. F., and Robillard, G. T. (1979a). *Escherichia coli* phosphoenolpyruvate dependent phosphotransferase system. Copurification of HPr and α1-6 glucan. *Biochemistry* **18**, 2990–2996.

Dooijewaard, G., Roossien, F. F., and Robillard, G. T. (1979b). *Escherichia coli* phosphoenolpyruvate dependent phosphotransferase system. NMR studies of the conformation of HPr and P-HPr and the mechanism of energy coupling. *Biochemistry* **18**, 2996–3001.

Ehring, R., Beyreuther, K., Wright, J. K., and Overath, P. (1980). *In vitro* and *in vivo* products of *E. coli* lactose permease gene are identical. *Nature (London)* **283**, 537–540.

El-Maghrabi, M. R., Claus, T. H., Pilkis, J., Fox, E., and Pilkis, S. J. (1982). Regulation of rat liver fructose 2,6-bisphosphatase. *J. Biol. Chem.* **257**, 7603–7607.

Enami, M., and Ishihama, A. (1984). Protein phosphorylation in *Escherichia coli* and purification of a protein kinase. *J. Biol. Chem.* **259**, 526–533.

Erni, B., Trachsel, H., Postma, P. W., and Rosenbusch, J. P. (1982). Bacterial phosphotransferase system. Solubilization and purification of the glucose-specific enzyme II from membranes of *Salmonella typhimurium. J. Biol. Chem.* **257**, 13726–13730.

Ferenci, T., and Boos, W. (1980). The role of the *Escherichia coli* lambda receptor in the transport of maltose and maltodextrins. *J. Supramol. Struct.* **13**, 101–116.

Ferenci, T., Schwentorat, M., Ullrich, S., and Vilmart, J. (1980). Lambda receptor in the outer membrane of *Escherichia coli* as a binding protein for maltodextrins and starch polysaccharides. *J. Bacteriol.* **142**, 521–526.

Feucht, B. U., and Saier, M. H., Jr. (1980). Fine control of adenylate cyclase by the phosphoenolpyruvate:sugar phosphotransferase system in *Escherichia coli* and *Salmonella typhimurium. J. Bacteriol.* **141**, 603–610.

Foster, D. L., Boublik, M., and Kaback, H. R. (1983). Structure of the *lac* carrier protein of *Escherichia coli. J. Biol. Chem.* **258**, 31–34.

Fowler, A. V., and Zabin, I. (1982). Sequence studies on the maltose-binding protein of *Escherichia coli. Ann. Microbiol. (Paris)* **133A**, 49–54.

Fraser, A. D. E., and Yamazaki, H. (1978). Construction of an *Escherichia coli* strain which excretes abnormally large amounts of adenosine 3′,5′-cyclic monophosphate. *Can. J. Microbiol.* **24**, 1423–1425.

Fraser, A. D. E., and Yamazaki, H. (1979). Effect of carbon sources on the rates of cyclic AMP synthesis, excretion, and degradation, and the ability to produce β-galactosidase in *Escherichia coli. Can. J. Biochem.* **57**, 1073–1079.

Fraser, A. D. E., and Yamazaki, H. (1983). Difference between glucose inhibition of glycerol and lactose utilization in *Escherichia coli. FEMS Microbiol. Lett.* **16**, 195–198.

Freundlieb, S., and Boos, W. (1982). Maltose transacetylase of *Escherichia coli:* A preliminary report. *Ann. Microbiol. (Paris)* **133A**, 181–190.

Fukada, H., Sturtevant, J. M., and Quiocho, F. A. (1983). Thermodynamics of the binding of L-arbinose and of D-galactose to the L-arabinose-binding protein of *Escherichia coli. J. Biol. Chem.* **258**, 13193–13196.

Gabay, J. (1982). Monoclonal antibodies against the *lamB* protein. *Ann. Microbiol. (Paris)* **133A**, 33–36.

Gabay, J., Benson, S., and Schwartz, M. (1983). Genetic mapping of antigenic determinants on a membrane protein. *J. Biol. Chem.* **258**, 2410–2414.

Garavito, R. M., Jenkins, J. A., Neuhaus, J. M., Pugsley, A. P., and Rosenbusch, J. P. (1982). Structural investigations of outer membrane proteins from *Escherichia coli. Ann. Mircrobiol. (Paris)* **133A**, 37–42.

Garcia, M. L., Patel, L., Padan, E., and Kaback, H. R. (1982). Mechanism of lactose transport in *Escherichia coli* membrane vesicles: Evidence for the involvement of histidine residue(s) in the response of the *lac* carrier to the proton electrochemical gradient. *Biochemistry* **21**, 5800–5805.

Garcia, M. L., Viitanen, P., Foster, D. L., and Kaback, H. R. (1983). Mechanism of lactose translocation in proteoliposomes reconstituted with *lac* carrier protein purified from *Escherichia coli*. I. Effect of pH and imposed membrane potential on efflux, exchange, and counterflow. *Biochemistry* **22**, 2524–2531.

Garnak, M., and Reeves, H. C. (1979). Purification and properties of phosphorylated isocitrate dehydrogenase of *Escherichia coli. J. Biol. Chem.* **254**, 7915–7920.

Gilliland, G. L., and Quiocho, F. A. (1981). Structure of the L-arabinose-binding protein from *Escherichia coli* at 2.4Å resolution. *J. Mol. Biol.* **146**, 341–362.

Gilson, E., Higgins, C. F., Hofnung, M., Ames, G. F. L., and Nikaido, H. (1982a). Extensive homology between membrane-associated components of histidine and maltose transport systems of *Salmonella typhimurium* and *Escherichia coli. J. Biol. Chem.* **257**, 9915–9918.

Gilson, E., Nikaido, H., and Hofnung, M. (1982b). Sequence of the *malK* gene in *E. coli* K12. *Nucleic Acids Res.* **10**, 7449–7458.

Glaser, J. H., and Conrad, H. E. (1980). Multiple kinetic forms of β-glucuronidase. *J. Biol. Chem.* **255**, 1879–1884.

Glesyna, M. L., Bol'shakova, T. N., and Gershanovich, V. N. (1983). Effect of *ptsI* and *ptsH* mutations on initiation of transcription of the *Escherichia coli* lactose operon. *Mol. Gen. Genet.* **190**, 417-420.

Goldenbaum, P. E., and Hall, G. A. (1979). Transport of cyclic adenosine 3'5'-monophosphate across *Escherichia coli* vesicle membranes. *J. Bacteriol.* **140**, 459–467.

Goldkorn, T., Simon, G., and Kaback, H. R. (1983). Topology of the *lac* carrier protein in the membrane of *Escherichia coli. Proc. Natl. Acad. Sci. U.S.A.* **80**, 3322–3326.

Goldkorn, T., Rimon, G., Kempner, E. S., and Kaback, H. R. (1984). Functional molecular weight of the *lac* carrier protein from *Escherichia coli* as studied by radiation inactivation analysis. *Proc. Natl. Acad. Sci. U.S.A.* **81**, 1021–1025.

Gonzalez, J. E., and Peterkofsky, A. (1977). The mechanism of sugar-dependent repression of synthesis of catabolic enzymes in *Escherichia coli. J. Supramol. Struct.* **6**, 495–502.

Greenwood, F. C., Hunter, W. M., and Glover, J. S. (1963). The preparation of ^{131}I-labeled human growth hormone of high specific radioactivity. *Biochem. J.* **89**, 114–123.

Grenier, F. C., Hayward, I., Novotny, M. J., Leonard, J. E., and Saier, M. H., Jr. (1985). Identification of enzyme III[gut]: An essential component of the glucitol phosphotransferase system in *Salmonella typhimurium. J. Bacteriol.* (in press).

Griesser, H. W., Müller-Hill, B., and Overath, P. (1983). Characterization of β-galactosidase-lactose-permease chimaeras of *Escherichia coli. Eur. J. Biochem.* **137**, 567–572.

Grill, H., Weigel, N., Gaffney, B. J., and Roseman, S. (1982). Sugar transport by the bacterial phosphotransferase system. Radioactive and electron paramagnetic resonance labeling of the *Salmonella typhimurium* phosphocarrier protein (HPr) at the NH₂-terminal methionine. *J. Biol. Chem.* **257**, 14510–14517.

Grossman, A. D., Ullmann, A., Burgess, R. R., and Gross, C. A. (1984). Regulation of cyclic AMP synthesis in *Escherichia coli* K-12: Effects of the *rpoD800* sigma mutation, glucose and chloramphenicol. *J. Bacteriol.* **158**, 110–114.

Guidi-Rontani, C., Danchin, A., and Ullmann, A. (1980). Catabolite repression in *Escherichia coli* mutants lacking cyclic AMP receptor protein. *Proc. Natl. Acad. Sci. U.S.A.* **77**, 5799–5801.

Guidi-Rontani, C., Danchin, A., and Ullmann, A. (1982). Pleiotropic expression of catabolic operons in the absence of the cAMP-CAP complex: The case of the maltose regulon. *Ann. Microbiol. (Paris)* **133A**, 81–86.

Gutnick, D., Calvo, J. M., Klopotowski, T., and Ames, B. N. (1969). Compounds which serve as the sole source of carbon or nitrogen for *Salmonella typhimurium* LT-2. *J. Bacteriol.* **100**, 215–219.

Haguenauer, R., and Kepes, A. (1971). The cycle of renewal of intracellular α-methyl glucoside accumulated by the glucose permease of *E. coli*. *Biochimie* **53**, 99–107.

Harwood, J. P., and Peterkofsky, A. (1975). Glucose-sensitive adenylate cyclase in toluene-treated cells of *Escherichia coli*. *J. Biol. Chem.* **250**, 4656–4662.

Harwood, J. P. Gazdar, C., Prasad, C., and Peterkofsky, A. (1976). Involvement of the glucose enzymes II of the sugar phosphotransferase system in the regulation of adenylate cyclase by glucose in *Escherichia coli*. *J. Biol. Chem.* **251**, 2462–2468.

Hatfield, D., Hofnung, M., and Schwartz, M. (1969). Nonsense mutations in the maltose A region of the genetic map of *Escherichia coli*. *J. Bacteriol.* **100**, 1311–1315.

Hayashi, S., and Lin, E. C. C. (1965). Product induction of glycerol kinase in *Escherichia coli*. *J. Mol. Biol.* **14**, 515–521.

Heller, K. B., Lin, E. C. C., and Wilson, T. H. (1980). Substrate specificity and transport properties of the glycerol facilitator of *Escherichia coli*. *J. Bacteriol.* **144**, 274–278.

Henderson, R. (1975). The structure of the purple membrane from *Halobacterium holobium*: Analysis of the X-ray diffraction pattern. *J. Mol. Biol.* **93**, 123–138.

Henderson, R., and Unwin, P. N. T. (1975). Three-dimensional model of purple membrane obtained by electron microscopy. *Nature (London)* **257**, 28–32.

Hengge, R. and Boos, W. (1983). Maltose and lactose transport in *Escherichia coli*. Examples of two different types of concentrative transport systems. *Biochim. Biophys Acta* **737**, 443–478.

Heuzenroeder, M. W., and Reeves, P. (1980). Periplasmic maltose-binding protein confers specificity on the outer membrane maltose pore of *Escherichia coli*. *J. Bacteriol.* **141**, 431–435.

Higgins, C. F., Haag, P. D., Nikaido, K., Ardeshir, F., Garcia, G., and Ferro-Luzzi Ames, G. (1982). Complete nucleotide sequence and identification of membrane components of the histidine transport operon of *S. typhimurium*. *Nature (London)* **298**, 723–727.

Hildenbrand, K., Brand, L., and Roseman, S. (1982). Sugar transport by the bacterial phosphotransferase system. Nanosecond fluorescence studies of the phospho-carrier protein (HPr) labeled at the NH_2-terminal methionine. *J. Biol. Chem.* **257**, 14518–14525.

Hoffee, P., and Englesberg, E. (1962). Effect of metabolic activity on the glucose permease of bacterial cells. *Proc. Natl. Acad. Sci. U.S.A.* **48**, 1759–1765.

Hofnung, M. (1982). Presentation of the maltose system and of the workshop. *Ann. Microbiol. (Paris)* **133A**, 5–9.

Hoving, H., Lolkema, J. S., and Robillard, G. T. (1981). *Escherichia coli* phos-phoenolpyruvate-dependent phosphotransferase system: Equilibrium kinetics and mechanism of enzyme I phosphorylation. *Biochemistry* **20**, 87–93.

Hoving, H., Koning, J. H., and Robillard, G. T. (1982). *Escherichia coli* phos-phoenolpyruvate-dependent phosphotransferase system: Role of divalent metals in the dimerization and phosphorylation of enzyme I. *Biochemistry* **21**, 3128–3135.

Hoving, H., Nowak, T., and Robillard, G. T. (1983). Escherichia coli phos-phoenolpyruvate dependent phosphotransferase system: Stereospecificity of proton transfer in the phosphorylation of enzyme I from Z-phosphoenol-butyrate. *Biochemistry* **21**, 3128–3136.

Huang, C. Y., Rhee, S. G., and Chock, P. B. (1982). Subunit cooperation and enzy-matic catalysis. *Annu. Revs. Biochem.* **51**, 935–971.

Hüdig, H., and Hengstenberg, W. (1980). The bacterial phosphoenolpyruvate depen-dent phosphotransferase system (PTS) Solubilisation and kinetic parameters of the glucose-specific membrane-bound enzyme II component of *Streptococcus faecalis*. *FEBS Lett.*, **114**, 103–106.

Hugenholtz, J., Hong, J. S., and Kaback, H. R. (1981). ATP-driven active transport in right-side-out bacterial membrane vesicles. *Proc. Natl. Acad. Sci. U.S.A.* **78**, 3446–3449.

Hunt, A. G., and Hong, J. S. (1983). Properties and characterization of binding-protein dependent active transport of glutamine in isolated membrane vesicles of *Escherichia coli*. *Biochemistry* **22**, 844–850.

Hunt, J. C., and Phibbs, P. V., Jr. (1981). Failure of Pseudomonas aeruginosa to form membrane-associated glucose dehydrogenase activity during anaerobic growth with nitrate. *Biochem. Biophys. Res. Commun.* **102**, 1393–1399.

Hunt, J. C., and Phibbs, P. V., Jr. (1983). Regulation of alternate peripheral pathways of glucose catabolism during aerobic and anaerobic growth of *Pseudomonas aeruginosa*. *J. Bacteriol.* **154**, 793–802.

Jablonski, E. G., Brand, L., and Roseman, S. (1983). Sugar transport by the bacterial phosphotransferase system. Preparation of a fluorescein derivative of the glu-cose specific phosphocarrier protein III[glc] and its binding to the phosphocarrier protein HPr. *J. Biol. Chem.* **258**, 9690–9699.

Jacobson, G. R., Lee, C. A., and Saier, M. H., Jr. (1979). Purification of the manni-tol-specific Enzyme II of the *Escherichia coli* phosphoenolpyruvate : sugar phosphotransferase system. *J. Biol. Chem.* **254**, 249–252.

Jacobson, G. R., Kelly, D. M., and Finlay, D. R. (1983a). The intramembrane topog-raphy of the mannitol-specific Enzyme II of the *Escherichia coli* phosphotrans-ferase system. *J. Biol. Chem.* **258**, 2955–2959.

Jacobson, G. R., Lee, C. A., Leonard, J. E., and Saier, M. H., Jr. (1983b). Mannitol-specific Enzyme II of the bacterial phosphotransferase system. I. Properties of the purified permease. *J. Biol. Chem.* **258**, 10748–10756.

Jacobson, G. R., Tanney, L. E., Kelly, D. M., Palman, K. B., and Corn, S. B. (1984). Substrate and phospholipid specificity of the purified mannitol permease of *Escherichia coli. J. Cell. Biochem.* **23**, 231–240.

Jin, R. Z., and Lin, E. C. C. (1984). An inducible phosphoenolpyruvate:dihydroxyacetone phosphotransferase system in *Escherichia coli. J. Gen. Microbiol.* **130**, 83–88.

Judewicz, N. D., De Robertis, E. M., Jr., and Torres, H. N. (1973). Inhibition of *Escherichia coli* growth by cyclic adenosine 3′,5′-monophosphate. *Biochem. Biophys. Res. Commun.* **52**, 1257–1262.

Kaback, H. R. (1983). The *lac* carrier protein in *Escherichia coli. J. Membr. Biol.* **76**, 95–112.

Kaczorowski, G. J., and Kaback, H. R. (1979). Mechanism of lactose translocation in membrane vesicles from *Escherichia coli.* I. Effect of pII on efflux, exchange, and counterflow. *Biochemistry,* **18**, 3691–3697.

Kaczorowski, G. J., Robertson, D. E., and Kaback, H. R. (1979). Mechanism of lactose translocation in membrane vesicles from *Escherichia coli.* II. Effect of imposed Δ ψ, Δ pH and Δ μH⁺. *Biochemistry* **18**, 3697–3704.

Kaczorowski, G. J., LeBlanc, G., and Kaback, H. R. (1980). Specific labeling of the *lac* carrier protein in membrane vesicles of *Escherichia coli* by a photoaffinity reagent. *Proc. Natl. Acad. Sci. U.S.A.* **77**, 6319–6323.

Kadner, R. J., and Shattuck-Eidens, D. M. (1983). Genetic control of hexose phosphate transport system of *Escherichia coli:* Mapping of deletion and insertion mutations in the *uhp* region. *J. Bacteriol.* **155**, 1052–1061.

Hunt, J. C., and Phibbs, P. V., Jr. (1981). Failure of *Pseudomonas aeruginosa* to form membrane-associated glucose dehydrogenase activity during anaerobic growth with nitrate. *Biochem. Biophys. Res. Commun.* **102**, 1393–1399.

Kay, W. W., and Cameron, M. (1978). Citrate transport in *Salmonella typhimurium. Arch. Biochem. Biophys.* **190**, 270–280.

Kennedy, E. P., Rumley, M. K., and Armstrong, J. B. (1974). Direct measurement of the binding of labeled sugars to the lactose permease M protein. *J. Biol. Chem.* **249**, 33–37.

Kier, L. O., Weppelman, R., and Ames, B. N. (1977a). Resolution and purification of three periplasmic phosphatases of *Salmonella typhimurium. J. Bacteriol.* **130**, 399–410.

Kier, L. O., Weppelman, R., and Ames, B. N. (1977b). Regulation of two phosphatases and a cyclic phosphodiesterase of *Salmonella typhimurium. J. Bacteriol.* **130**, 420–428.

Knowles, J. R. (1980). Enzyme-catalyzed phosphoryl transfer reactions. *Annu. Rev. Biochem.* **49**, 877–919.

Konings, W. N., and Robillard, G. T. (1982). Physical mechanism for regulation of proton solute symport in *Escherichia coli. Proc. Natl. Acad. Sci. U.S.A.* **79**, 5480–5484.

Kornberg, H. L. (1973). Fine control of sugar uptake by *Escherichia coli. Symp. Soc. Exp. Biol.* **27**, 175–193.

Kornberg, H. L., and Reeves, R. E. (1972). Inducible phosphoenolpyruvate-dependent hexose phosphotransferase activities in *Escherichia coli. Biochem. J.* **128**, 1339–1344.

Kornberg, H. L., and Watts, P. D. (1978). Roles of *crr* gene products in regulating carbohydrate uptake by *Escherichia coli. FEBS Lett.* **89**, 329–339.

Krebs, E. G., and Beavo, J. A. (1979). Phosphorylation-dephosphorylation of enzymes. *Annu. Rev. Biochem.* **48**, 923–959.

Kubota, Y., Iuchi, S., Fujisawa, A., and Tanaka, S. (1979). Separation of four components of the phosphoenolpyruvate : glucose phosphotransferase system in *Vibrio parahaemolyticus. Microbiol. Immunol.* **23**, 131–146.

Kukuruzinska, M. A., Harrington, W. F., and Roseman, S. (1982). Sugar transport by the bacterial phosphotransferase system. Studies on the molecular weight and association of enzyme I. *J. Biol. Chem.* **257**, 14470–14476.

Kundig, W., and Roseman, S. (1971). Sugar transport. Characterization of constitutive membrane-bound enzymes II of the *Escherichia coli* phosphotransferase systems. *J. Biol. Chem.* **246**, 1407–1418.

Kustu, S. G., McFarland, N. C., Hui, S. P., Esmon, B., and Ferro-Luzzi Ames, G. (1979). Nitrogen control in *Salmonella typhimurium*: co-regulation of synthesis of glutamine synthetase and amino acid transport systems. *J. Bacteriol.* **138**, 218–234.

Kyte, J., and Doolittle, R. F. (1982). A simple method for displaying the hydropathic character of a protein. *J. Mol. Biol.* **157**, 105–132.

Lancaster, J. R., Jr. (1982). *Hypothesis*—Mechanism of lactose–proton cotransport in *Escherichia coli:* Kinetic results in terms of the site exposure model. *FEBS Lett.* **150**, 9–18.

La Porte, D. C., and Koshland, D. E., Jr. (1982). A protein with kinase and phosphatase activities involved in regulation of tricarboxylic acid cycle. *Nature (London)* **300**, 458–460.

Lee, C. A., and Saier, M. H., Jr. (1983a). Use of cloned *mtl* genes of *Escherichia coli* to introduce *mtl* deletion mutations into the chromosome. *J. Bacteriol.* **153**, 685–692.

Lee, C. A., and Saier, M. H., Jr. (1983b). Mannitol-specific enzyme II of the bacterial phosphotransferase system. III. The nucleotide sequence of the permease gene. *J. Biol. Chem.* **258**, 10761–10767.

Lee, C. A., Jacobson, G. R., and Saier, M. H., Jr. (1981). Plasmid-directed synthesis of enzymes required for D-mannitol transport and utilization in *Escherichia coli. Proc. Natl. Acad. Sci. U.S.A.* **78**, 7336–7340.

Lee, L. G., Britton, P., Parra, F., Boronat, A., and Kornberg, H. (1982). Expression of the *ptsH*+ gene of *Escherichia coli* cloned on plasmid pBR322. *FEBS Lett.* **149**, 288–292.

Lengeler, J. (1975a). Mutations affecting transport of the hexitols, D-mannitol, D-glucitol and D-galactitol in *Escherichia coli* K-12: isolation and mapping. *J. Bacteriol.* **124**, 26–38.

Lengeler, J. (1975b). Nature and properties of hexitol transport systems in *Escherichia coli. J. Bacteriol.* **124**, 39–47.

Lengeler, J. (1977). Analysis of mutations affecting the dissimilation of galactitol (dulcitol) in *Escherichia coli* K 12. *Mol. Gen. Genet.* **152**, 83–91.

Lengeler, J., and Steinberger, H. (1978a). Analysis of the regulatory mechanisms controlling the synthesis of the hexitol transport systems in *Escherichia coli* K12. *Mol. Gen. Genet.* **164**, 163–169.

Lengeler, J., and Steinberger, H. (1978b). Analysis of regulatory mechanisms controlling the activity of hexitol transport systems in *Escherichia coli* K12. *Mol. Gen. Genet.* **167**, 75–82.

Lengeler, J., Auburger, A.-M., Mayer, R., and Pecher, A. (1981). The phosphoenolpyruvate-dependent carbohydrate: Phosphotransferase system enzymes II as chemoreceptors in chemotaxis of *Escherichia coli* K 12. *Mol. Gen. Genet.* **183**, 163–170.

Lengeler, J. W., Mayer, R. J., and Schmid, K. (1982). Phosphoenolpyruvate-dependent phosphotransferase system Enzyme III and plasmid-encoded sucrose transport in *Escherichia coli* K-12. *J. Bacteriol.* **151**, 468–471.

Leonard, J. E., and Saier, M. H., Jr. (1981). Genetic dissection of catalytic activities of the *Salmonella typhimurium* mannitol enzyme II. *J. Bacteriol.* **145**, 1106–1109

Leonard, J. E., and Saier, M. H., Jr, (1983). Mannitol-specific Enzyme II of the bacterial phosphotransferase system. II. Reconstitution of vectorial transphosphorylation in phospholipid vesicles. *J. Biol. Chem.* **258**, 10757–10760.

Leonard, J. E., Lee, C. A., Apperson, A. J., Dills, S. S., and Saier, M. H., Jr. (1981). The role of membranes in the transport of small molecules. *In* "Organization of Prokaryotic Cell Membranes" (B. K. Gosh, ed.), Vol. 1, pp. 1–52.

Levinthal, M. (1971). Biochemical studies of melibiose metabolism in wild type and *mel* mutant strains of *Salmonella typhimurium. J. Bacteriol.* **105**, 1047–1052.

Lin, E. C. C. (1970). The genetics of bacterial transport systems. *Annu. Rev. Genet.* **4**, 225–262.

Lin, E. C. C. (1976). Glycerol dissimilation and its regulation in bacteria. *Annu. Rev. Microbiol.* **30**, 535–578.

Lin, E. C. C. (1977). Glycerol utilization and its regulation in mammals. *Annu. Rev. Biochem.* **46**, 765–795.

Lo, T. C. Y. (1977). The molecular mechanism of dicarboxylic acid transport in *Escherichia coli* K 12. *J. Supramol. Struct.* **7**, 463–480.

Lo, T. C. Y. (1979). The molecular mechanisms of substrate transport in Gram-negative bacteria. *Can. J. Biochem.* **57**, 289–301.

Lo, T. C. Y. (1981). Use of a nonpenetrating substrate analogue to study the molecular mechanism of the outer membrane dicarboxylate transport system in *Escherichia coli* K12. *J. Biol. Chem.* **256**, 5511–5517.

Lo, T. C. Y., and Bewick, M. A. (1981). Use of a nonpenetrating substrate analogue to study the molecular mechanism of the outer membrane dicarboxylate transport system in *Escherichia coli* K12. *J. Biol. Chem.* **256**, 5511–5517.

Lombardi, F. J. (1981). Lactose-H$^+$($^-$OH)transport system of *Escherichia coli*: Multistate gated pore model based on half-sites stoichiometry for high-affinity substrate binding in a symmetrical dimer. *Biochim. Biophys. Acta* **649**, 661–679.

Lombardi, F. J., and Kaback, H. R. (1972). Mechanisms of active transport in isolated bacterial membrane vesicles. VIII. The transport of amino acids by membranes prepared from *Escherichia coli. J. Biol. Chem.* **247**, 7844–7857.

London, J., and Chace, N. M. (1977). New pathway for the metabolism of pentitols. *Proc. Natl. Acad. Sci. U.S.A.* **74**, 4296–4300.

London, J., and Chace, N. M. (1979). Pentitol metabolism in *Lactobacillus casei. J. Bacteriol.* **140**, 949–954.

London, J., and Hausman, S. (1982). Xylitol-mediated transient inhibition of ribitol utilization by *Lactobacillus casei. J. Bacteriol.* **150**, 657–661.

London, J., and Hausman, S. (1983). Purification and characterization of the III^{xyl} phospho-carrier protein of the phosphoenolpyruvate-dependent xylitol: Phosphotransferase found in *Lactobacillus casei* C183. *J. Bacteriol.* **156**, 611–619.

Long, M. M., Urry, D. W., and Stoeckenius, W. (1977). Circular dichroism of biological membranes: Purple membrane of *Halobacterium halobium*. *Biochem. Biophys. Res. Commun.* **75**, 725–731.

Luckey, M., and Nikaido, H. (1980a). Specificity of diffusion channels produced by the lambda phage receptor protein of *Escherichia coli*. *Proc. Natl. Acad. Sci. U.S.A.* **77**, 167–171.

Luckey, M., and Nikaido, H. (1980b). Diffusion of solutes through channels produced by phage lambda receptor protein of *Escherichia coli*: Inhibition by higher oligosaccharides of maltose series. *Biochem. Biophys. Res. Commun.* **93**, 166–171.

Luckey, M., and Nikaido, H. (1982). Purification and functional characterization of mutant *lamB* proteins. *Ann. Microbiol. (Paris)* **133A**, 165–166.

Luckey, M., and Nikaido, H. (1983). Bacteriophage lambda receptor protein in *Escherichia coli* K-12: Lowered affinity of some mutant proteins for maltose-binding protein *in vitro*. *J. Bacteriol.* **153**, 1056–1059.

Lugtenberg, B., and Van Alphen, L. (1983). Molecular architecture and functioning of the outer membrane of *Escherichia coli* and other Gram-negative bacteria. *Biochim. Biophys. Acta* **737**, 51–115.

McGinnis, J. F., and Paigen, K. (1967). Catabolite inhibition: A general phenomenon in the control of carbohydrate utilization. *J. Bacteriol.* **100**, 902–913.

McGinnis, J. F., and Paigen, K. (1973). Site of catabolite inhibition of carbohydrate metabolism. *J. Bacteriol.* **114**, 885–887.

McKay, D. B., and Steitz, T. A. (1981). Structure of catabolite gene activator protein at 2.9 Å resolution suggests binding to left-handed B-DNA. *Nature (London)* **290**, 744–749.

Magasanik, B. (1970). Glucose effects: Inducer exclusion and repression. *In* "The Lactose Operon" (J. R. Beckwith and D. Zipser, eds.), pp. 189–219. Cold Spring Harbor Lab., Cold Spring Harbor, New York.

Majerfeld, I. H., Miller, D., Spitz, E., and Rickenberg, H. V. (1981). Regulation of the synthesis of adenylate cyclase in *Escherichia coli* by the cAMP-cAMP receptor protein complex. *Mol. Gen. Genet.* **181**, 470–475.

Makman, R. S., and Sutherland, E. W. (1965). Adenosine 3',5'-phosphate in *Escherichia coli*. *J. Biol. Chem.* **240**, 1309–1314.

Malloy, P. J., and Reeves, H. C. (1983). Cyclic AMP-independent phosphorylation of *Escherichia coli* isocitrate dehydrogenase. *FEBS Lett.* **151**, 59–62.

Manai, M., and Cozzone, A. J. (1983). Characterization of the amino acids phosphorylated in *E. coli* proteins. *FEMS Microbiol. Lett.* **17**, 87–91.

Marechal, R. (1984). Transport and metabolism of trehalose in *Escherichia coli* and *Salmonella typhimurium*. *Arch. Microbiol.* **137**, 70–73.

Matsushita, K., Patel, L., Gennis, R. B., and Kaback, H. R. (1983). Reconstitution of active transport in proteoliposomes containing cytochrome *o* oxidase and *lac* carrier protein purified from *Escherichia coli*. *Proc. Natl. Acad. Sci. U.S.A.* **80**, 4889–4893.

Meadow, N. D., and Roseman, S. (1982). Sugar transport by the bacterial phosphotransferase system. Isolation and characterization of a glucose-specific phosphocarrier protein (III[Glc]) from *Salmonella typhimurium*. *J. Biol. Chem.* **257**, 14526–14537.

Meadow, N. D., Rosenberg, J. M., Pinkert, H. M., and Roseman, S. (1982a). Sugar transport by the bacterial phosphotransferase system. Evidence that *crr* is the structural gene for the *Salmonella typhimurium* glucose-specific phosphocarrier protein III^{Glc}. *J. Biol. Chem.* **257**, 14538–14542.

Meadow, N. D., Saffen, D. W., Dottin, R. P., and Roseman, S. (1982b). Molecular cloning of the *crr* gene and evidence that it is the structural gene for III^{glc}, a phosphocarrier protein of the bacterial phosphotransferase system. *Proc. Natl. Acad. Sci. U.S.A.* **79**, 2528–2532.

Mieschendahl, M., Buchel, D., Böcklage, H., and Müller-Hill, B. (1981). Mutations in the *lacY* gene of *Escherichia coli* define functional organization of lactose permease. *Proc. Natl. Acad. Sci. U.S.A.* **78**, 7652–7656.

Miller, D. M., III, Olson, J. S., Pflugrath, J. W., and Quiocho, F. A. (1983). Rates of ligand binding to periplasmic proteins involved in bacterial transport and chemotaxis. *J. Biol. Chem.* **258**, 13665–13672.

Misset, O., and Robillard, G. T. (1982). *Escherichia coli* phosphoenolpyruvate-dependent phosphotransferase system: Mechanism of phosphoryl-group transfer from phosphoenolpyruvate to HPr. *Biochemistry* **21**, 3136–3142.

Misset, O., Brouwer, M., and Robillard, G. T. (1980). *Escherichia coli* phosphoenolpyruvate-dependent phosphotransferase system. Evidence that the dimer is the active form of Enzyme I. *Biochemistry* **19**, 883–890.

Mitchell, P. (1963). Molecule, group, and electron translocation through natural membranes. *Biochem. Soc. Symp.* **22**, 142–168.

Mitchell, P. (1967). Translocation through natural membranes. *Adv. Enzymol.* **29**, 33–87.

Mitchell, P. (1979). The Ninth Sir Hans Krebs Lecture. Compartmentation in living systems. Ligand conduction: A general catalytic principle in chemical, osmotic, and chemiosmotic reaction systems. *Eur. J. Biochem.* **95**, 1–20.

Mitchell, W. J., Misko, T. P., and Roseman, S. (1982). Sugar transport by the bacterial phosphotransferase system. Regulation of other transport systems (lactose and melibiose). *J. Biol. Chem.* **257**, 14553–14564.

Monard, D., Janecek, J., and Rickenberg, H. V. (1969). The enzymic degradation of 3',5' cyclic AMP in strains of *E. coli* sensitive and resistant to catabolite repression. *Biochem. Biophys. Res. Commun.* **35**, 584–591.

Nakai, T., and Ishii, J. N. (1982). Molecular weights and subunit structure of *lamB* proteins. *Ann. Microbiol. (Paris)* **133A**, 21–26.

Nelson, S. O., and Postma, P. W. (1984). Interactions *in vivo* between III^{Glc} of the phosphoenolpyruvate : sugar phosphotransferase system and the glycerol and maltose uptake systems of *Salmonella typhimurium*. *Eur. J. Biochem.* **139**, 29–34.

Nelson, S. O., Scholte, B. J., and Postma, P. W. (1982). Phosphoenolpyruvate : sugar phosphotransferase system-mediated regulation of carbohydrate metabolism in *Salmonella typhimurium*. *J. Bacteriol.* **150**, 604–615.

Nelson, S. O., Wright, J. K., and Postma, P. W. (1983). The mechanism of inducer exclusion. Direct interaction between purified III^{glc} of the phosphoenolypyruvate : sugar phosphotransferase system and the lactose carrier of *Escherichia coli*. *EMBO J.* **2**, 715–720.

Nelson, S. O., Lengeler, J., and Postma, P. W. (1984). The role of III^{Glc} of the PEP: glucose phosphotransferase system in inducer exclusion in *Escherichia coli*. *J. Bacteriol.* **160**, 360–364.

Neuhaus, J.-M. (1982). The receptor protein of phage lambda: Purification, characterization and preliminary electrical studies in planar lipid bilayers. *Ann. Microbiol.* (*Paris*) **133A,** 27–32.

Neuhaus, J.-M. Schindler, H., and Rosenbusch, J. P. (1983). The periplasmic maltose binding protein modifies the channel-characteristics of maltoporin. *EMBO J.* **2,** 1987–1992.

Newcomer, M. E., Miller, D. M., and Quiocho, F. A. (1979). Location of the sugar-binding site of L-arabinose-binding protein. *J. Biol. Chem.* **254,** 7529–7533.

Newcomer, M. E., Gilliland, G. L., and Quiocho, F. A. (1981a). L-Arabinose-binding protein–sugar complex at 2.4 Å resolution. *J. Biol. Chem.* **256,** 13213–13217.

Newcomer, M. E., Lewis, B. A., and Quiocho, F. A. (1981b). The radius of gyration of L-arabinose-binding protein decreases upon binding of ligand. *J. Biol. Chem.* **256,** 13218–13222.

Newman, M. J., and Wilson, T. H. (1980). Solubilization and reconstitution of the lactose transport system from *Escherichia coli. J. Biol. Chem.* **255,** 10583–10586.

Newman, M. J., Foster, D. L., Wilson, T. H., and Kaback, H. R. (1981). Purification and reconstitution of functional lactose carrier from *Escherichia coli. J. Biol. Chem.* **256,** 11804–11808.

Niiya, S., Yamasaki, K., Wilson, T. H., and Tsuchiya, T. (1982). Altered cation coupling to melibiose transport in mutants of *Escherichia coli. J. Biol. Chem.* **257,** 8902–8906.

Nikaido, H., and Wu, H. C. P. (1984). Amino acid sequence homology among the major outer membrane proteins of *Escherichia coli. Proc. Natl. Acad. Sci. U.S.A.* **81,** 1048–1052.

Nikaido, H., Luckey, M., and Rosenberg, E. Y. (1980). Non-specific and specific diffusion channels in the outer membrane of *Escherichia coli. J. Supramol. Struct.* **13,** 305–313.

Niwano, M., and Taylor, B. L. (1982). Novel sensory adaptation mechanism in bacterial chemotaxis to oxygen and phosphotransferase substrates. *Proc. Natl. Acad. Sci. U.S.A.* **79,** 11–15.

Novotny, M. J., Frederickson, W. L., Waygood, E. B., and Saier, M. A., Jr., (1985). Allosteric regulation of glycerol kinase by Enzyme III^{glc} of the phosphotransferase system in *Escherichia coli* and *Salmonella typhimuvium. J. Bacteriol.* (in press).

Ohki, M., Ogawa, H., and Nishimura, S. (1982). Synthesis of mRNA ot *malB* operons at specific stages in the cell cycle of *Escherichia coli. Ann. Microbiol* (*Paris*) **133A,** 71–76.

Ohlendorf, D. H., Anderson, W. F., Fisher, R. G., Takeda, Y., and Matthews. B. W. (1982). The molecular basis of DNA-protein recognition inferred from the structure of cro repressor. *Nature* (*London*) **298,** 718–723.

O'Neill, M. C., Amass, K., and de Crombrugghe, B. (1981). Molecular model of the DNA interaction site for the cyclic AMP receptor protein. *Proc. Natl. Acad. Sci. U.S.A.* **78,** 2213–2217.

Osumi, T., and Saier, M. H., Jr. (1982a). Regulation of lactose permease activity by the phospho*enol*pyruvate : sugar phosphotransferase system: Evidence for direct binding of the glucose-specific enzyme III to the lactose permease. *Proc. Natl. Acad. Sci. U.S.A.* **79,** 1457–1461.

Osumi, T., and Saier, M. H., Jr. (1982b). Mechanism of regulation of the lactose permease by the phosphotransferase system in *Escherichia coli*: Evidence for protein-protein interaction. *Ann. Microbiol. (Paris)* **133A**, 269–273.

Ovchinnikov, Y. A. (1982). Rhodopsin and bacteriorhodopsin: Structure–function relationships. *FEBS Lett.* **148**, 179–191.

Overath, P., and Wright, J. K. (1983). Lactose permease: A carrier on the move. *TIBS* **8**, 404–408.

Overath, P., Teather, R. M., Simoni, R. D., Aichele, G., and Wilhelm, U. (1979). Lactose carrier protein of *Escherichia coli* during the formation of spheroplasts. *Biochemistry* **18**, 1–11.

Page, M. G. P, and West, I. C. (1980) Kinetics of lactose transport into *Escherichia coli* in the presence and absence of a proton motive force. *FEBS Lett.* **120**, 187–191.

Page, M. G. P., and West, I. C. (1981). The kinetics of the β-galactoside-proton symport of *Escherichia coli*. *Biochem. J.* **196**, 721–731.

Paigen, K. (1966). Phenomenon of transient repression in *Escherichia coli*. *J. Bacteriol.* **91**, 1201–1209.

Parra, F., Jones-Mortimer, M. C., and Kornberg, H. L. (1983). Phosphotransferase-mediated regulation of carbohydrate utilization in *Escherichia coli*. *J. Gen. Microbiol.* **129**, 337–348.

Patel, L., Garcia, M. L., and Kaback, H. R. (1982). Direct measurement of lactose/proton symport in *Escherichia coli* membrane vesicles: Further evidence for the involvement of histidine residue(s). *Biochemistry* **21**, 5805–5810.

Pecher, A., Renner, I., and Lengeler, J. W. (1983). The phosphoenolpyruvate-dependent carbohydrate : phosphotransferase system enzymes II, a new class of chemosensors in bacterial chemotaxis. *In* "Mobility and Recognition in Cell Biology" (H. Sund and C. Veeger, eds.), pp. 517–531. de Gruyter, Berlin.

Perret, J., and Gay, P. (1979). Kinetic study of a phosphoryl exchange reaction between fructose and fructose-1-phosphate catalyzed by the membrane-bound Enzyme II of the phospho*enol* pyruvate-fructose 1-phospho-transferase system of *Bacillus subtilis*. *Eur. J. Biochem.* **102**, 237–246.

Peterkofsky, A., and Gazdar, C. (1974). Glucose inhibition of adenylate cyclase in intact cells of *Escherichia coli*. *Proc. Natl. Acad. Sci. U.S.A.* **71**, 2324–2328.

Peterkofsky, A., and Gazdar, C. (1975). Interaction of enzyme I of the phospho*enol*pyruvate : sugar phosphotransferase system with adenylate cyclase of *Escherichia coli*. *Proc. Natl. Acad. Sci. U.S.A.* **72**, 2920–2924.

Peterkofsky, A., and Gazdar, C. (1978). The *Escherichia coli* adenylate cyclase complex: Activation by phosphoenolpyruvate. *J. Supramol. Struct.* **9**, 219–230.

Peterkofsky, A., and Gazdar, C. (1979). *Escherichia coli* adenylate cyclase complex: Regulation by the proton electrochemical gradient. *Proc. Natl. Acad. Sci. U.S.A.* **76**, 1099–1103.

Peterkofsky, A., Harwood, J., and Gazdar, C. (1975). Inducibility of sugar sensitivity of adenylate cyclase of *E. coli*. *B. J. Cyclic Nucleotide Res.* **1**, 11–20.

Postma, P. W. (1982). Regulation of sugar transport in *Salmonella typhimurium*. *Ann. Microbiol. (Paris)* **133A**, 261–267.

Postma, P. W., and Roseman, S. (1976). The bacterial phosphoenolpyruvate : sugar phosphotransferase system. *Biochim. Biophys. Acta* **457**, 213–257.

Postma, P. W., and Scholte, B. J. (1979). Regulation of sugar transport in *Salmonella typhimurium. In* "Function and Molecular Aspects of Biomembrane Transport" (E. Quagliariello *et al.,* eds.), pp. 249–257. Elsevier/North-Holland Biomedical Press, Amsterdam.

Postma, P. W., Epstein, W., Schuitema, A. R. J., and Nelson, S. O. (1984). Interaction between III^Glc of the phosphoenolpyruvate : sugar phosphotransferase system and glycerol kinase of *Salmonella typhimurium. J. Bacteriol.* **158,** 351–353.

Potter, K., Chaloner-Larsson, G., and Yamazaki, H. (1974). Abnormally high rate of cAMP excretion from an *Escherichia coli* mutant deficient in cyclic AMP receptor protein. *Biochem. Biophys. Res. Commun.* **57,** 379–385.

Price, V. L., and Gallant, J. A. (1983). The glucose effect in *Bacillus subtilis. Eur. J. Biochem.* **134,** 105–107.

Racker, E., Violand, B., O'Neal, S., Alfonzo, M., and Telford, J. (1979). Reconstitution, a way of biochemical research: Some new approaches to membrane bound enzymes. *Arch. Biochem. Biophys.* **198,** 470–477.

Reider, E., Wagner, E. F., and Schweiger, M. (1979). Control of phospho*enol*pyruvate-dependent phosphotransferase-mediated sugar transport in *Escherichia coli* by energization of the cell membrane. *Proc. Natl. Acad. Sci. U.S.A.* **76,** 5529–5533.

Reizer, J., and Panos, C. (1980). Regulation of β-galactoside phosphate accumulation in *Streptococcus pyogenes* by an expulsion mechanism. *Proc. Natl. Acad. Sci. U.S.A.* **77,** 5497–5501.

Reizer, J., and Saier, M. H., Jr. (1983). Evidence for the involvement of the lactose Enzyme II of the phosphotransferase system in the rapid expulsion of free galactosides from *Streptococcus pyogenes. J. Bacteriol.* **156,** 236–242.

Reizer, J., Novotny, M. J., Panos, C., and Saier, M. H., Jr. (1983). The mechanism of inducer expulsion in *Streptococcus pyogenes*: A two step process activated by ATP. *J. Bacteriol.* **156,** 354–361.

Reizer, J., Hengstenberg, W., Novotny, M. J., Grenier, F. C., and Saier, M. H., Jr. (1985). The phosphotransferase system and the regulation of sugar transport by protein phosphorylation in Gram-positive bacteria. *CRC Int. Rev. Microbiol.* (in preparation).

Rephaeli, A. W., and Saier, M. H., Jr. (1976). Effects of *crp* mutations on adenosine 3′,5′-monophosphate metabolism in *Salmonella typhimurium. J. Bacteriol.* **127,** 120–127.

Rephaeli, A. W., and Saier, M. H., Jr. (1978). Kinetic analyses of the sugar phosphate : sugar transphosphorylation reaction catalyzed by the glucose enzyme II complex of the bacterial phosphotransferase system. *J. Biol. Chem.* **253,** 7595–7597.

Rephaeli, A. W., and Saier, M. H., Jr. (1980a). Substrate specificity and kinetic characterization of sugar uptake and phosphorylation, catalyzed by the mannose enzyme II of the phosphotransferase system in *Salmonella typhimurium. J. Biol. Chem.* **255,** 8585–8591.

Rephaeli, A. W., and Saier, M. H., Jr. (1980b). Regulation of genes coding for enzyme constituents of the bacterial phosphotransferase system. *J. Bacteriol.* **141,** 658–663.

Rephaeli, A. W., Artenstein, I. R., and Saier, M. H., Jr. (1980). Physiological function of periplasmic hexose phosphatase in *Salmonella typhimurium. J. Bacteriol.* **141,** 1474–1477.

Rickenberg, H. V. (1974). Cyclic AMP in prokaryotes. *Annu. Rev. Microbiol.* **28**, 353–369.

Riley, M., and Anilionis, A. (1978). Evolution of the bacterial genome. *Annu. Rev. Microbiol.* **32**, 519–560.

Roberts, R. B., Abelson, P. H., Cowie, D. B., Bolton, E. I., and Britten, R. J. (1963). "Studies of Biosynthesis in *E. coli*," Carnegie Inst. Washington Publ. 607. Kirby Lithography Co., Washington, D.C.

Robertson, D. E., Kaczorowski, G. J., Garcia, M. L., and Kaback, H. R. (1980). Active transport in membrane vesicles from *Escherichia coli*: The electrochemical proton gradient alters the distribution of the *lac* carrier between two different kinetic states. *Biochemistry* **19**, 5692–5702.

Robillard, G. T. (1982). The enzymology of the bacterial phosphoenolpyruvate-dependent sugar transport systems. *Mol. Cell. Biochem.* **16**, 3–24.

Robillard, G. T., and Konings, W. N. (1981). Physical mechanism for regulation of phosphoenolpyruvate-dependent glucose transport activity in *Escherichia coli*. *Biochemistry* **20**, 5025–5032.

Robillard, G. T., and Lageveen, R. G. (1982). Non-vectorial phosphorylation by the bacterial PEP-dependent phosphotransferase system is an artifact of spheroplast and membrane vesicle preparation procedures. *FEBS Lett.* **147**, 143–148.

Robillard, G. T., Dooijewaard, G., and Lolkema, J. (1979). *Escherichia coli* phosphoenolpyruvate dependent phosphotransferase system. Complete purification of Enzyme I by hydrophobic interaction chromatography. *Biochemistry* **18**, 2984–2989.

Roossien, F. F., Dooijewaard, G., and Robillard, G. T. (1979). The *Escherichia coli* phosphoenolpyruvate-dependent phosphotransferase system: Observation of heterogeneity in the amino acid composition of HPr. *Biochemistry* **18**, 5794–5797.

Rosen, B. P., ed. (1978). "Bacterial Transport." Dekker, New York.

Rosenbusch, J. P. (1974). Characterization of the major envelope protein from *Escherichia coli*. *J. Biol. Chem.* **249**, 8019–8029.

Roy, A., and Danchin, A. (1982). The *cya* locus of *Escherichia coli* K12: Organization and gene products. *Mol. Gen. Genet.* **188**, 465–471.

Roy, A., Danchin, A., Joseph, E., and Ullmann, A. (1983a). Two functional domains in adenylate cyclase of *Escherichia coli*. *J. Mol. Biol.* **165**, 197–202.

Roy, A., Haziza, C., and Danchin, A. (1983b). Regulation of adenylate cyclase synthesis in *Escherichia coli*: Nucleotide sequence of the control region. *EMBO J.* **2**, 791–797.

Saier, M. H., Jr. (1977). Bacterial phosphoenolpyruvate:sugar phosphotransferase systems: Structural, functional, and evolutionary interrelationships. *Bacteriol. Rev.* **41**, 856–871.

Saier, M. H., Jr. (1979). The role of the cell surface in regulating the internal environment. *In* "The Bacteria" (I. C. Gunsalus and R. Y. Stanier, eds.), Vol. 7, pp. 167–227. Academic Press, New York.

Saier, M. H., Jr. (1980). Catalytic activities associated with the Enzymes II of the bacterial phosphotransferase system. *J. Supramol. Struct.* **14**, 281–294.

Saier, M. H., Jr. (1982). The bacterial phosphotransferase system in regulation of carbohydrate permease synthesis and activity. *In* "Membranes and Transport" (A. N. Martonosi, ed.), Vol. 2, pp. 27–32. Plenum, New York.

Saier, M. H., Jr., and Boyden, D. A. (1984). Mechanism, regulation and physiological significance of the loop diuretic-sensitive NaCl/KCl symport system in animal cells. *Mol. Cell. Biochem.* **59,** 11–32.

Saier, M.`H., Jr., and Feucht, B. U. (1975). Coordinate regulation of adenylate cyclase and carbohydrate permeases by the phosphoenolpyruvate:sugar phosphotransferase system in *Salmonella typhimurium. J. Biol. Chem.* **250,** 7078–7080.

Saier, M. H., Jr., and Jacobson, G. R. (1984). "The Molecular Basis of Sex and Differentiation. A Comparative Study of Evolution, Mechanism and Control in Microorganisms." Springer-Verlag, New York, (in press).

Saier, M. H., Jr., and Leonard, J. E. (1983). The mannitol enzyme II of the bacterial phospho-transferase system: A functionally chimaeric protein with receptor, transport, kinase and regulatory activities. *In* "Multifunctional Proteins: Catalytic/Structural and Regulatory" (J. F. Kane, Ed.), pp. 11–30. CRC Press, Boca Raton, Florida.

Saier, M. H., Jr., and Moczydlowski, E. G. (1978). The regulation of carbohydrate transport in *Escherichia coli* and *Salmonella typhimurium. In* "Bacterial Transport" (B. Rosen, ed.), pp. 103–122. Dekker, New York.

Saier, M. H., Jr., and Roseman, S. (1972). Inducer exclusion and repression of enzyme synthesis in mutants of *Salmonella typhimurium* defective in enzyme I of the phosphoenolpyruvate:sugar phosphotransferase system. *J. Biol. Chem.* **247,** 972–975.

Saier, M. H., Jr., and Roseman, S. (1976a). Sugar transport. The *crr* mutation: Its effect on repression of enzyme synthesis. *J. Biol. Chem.* **251,** 6598–6605.

Saier, M. H., Jr., and Roseman, S. (1976b). Sugar transport. Inducer exclusion and regulation of the melibiose, maltose, glycerol, and lactose transport systems by the phosphoenolpyruvate:sugar phosphotransferase system. *J. Biol. Chem.* **251,** 6606–6615.

Saier, M. H., Jr., and Schmidt, M. R. (1981). Vectorial and nonvectorial transphosphorylation catalyzed by enzymes II of the bacterial phosphotransferase system. *J. Bacteriol.* **145,** 391–397.

Saier, M. H., Jr., and Simoni, R. D. (1976). Regulation of carbohydrate uptake in Gram-positive bacteria. *J. Biol. Chem.* **251,** 893–894.

Saier, M. H., Jr., and Stiles, C. D. (1975). "Molecular Dynamics in Biological Membranes," Springer-Verlag, New York.

Saier, M. H., Jr., Simoni, R. D., and Roseman, S. (1970). The physiological behavior of enzyme I and heat-stable protein mutants of a bacterial phosphotransferase system. *J. Biol. Chem.* **254,** 5870–5873.

Saier, M. H., Jr., Feucht, B. U., and Roseman, S. (1971). Phosphoenolpyruvate-dependent fructose phosphorylation in photosynthetic bacteria. *J. Biol. Chem.* **246,** 7819–7821.

Saier, M. H., Jr., Wentzel, D. L., Feucht, B. U., and Judice, J. J. (1975a). A transport system for phosphoenolpyruvate, 2-phosphoglycerate, and 3-phosphoglycerate in *Salmonella typhimurium. J. Biol. Chem.* **250,** 5089–5096.

Saier, M. H., Jr., Feucht, B. U., and McCaman, M. T. (1975b). Regulation of intracellular adenosine cyclic 3′:5′-monophosphate levels in *Escherichia coli* and *Salmonella typhimurium. J. Biol. Chem.* **250,** 7593–7601.

Saier, M. H., Jr., Feucht, B. U., and Hofstadter, L. J. (1976a). Regulation of carbohydrate uptake and adenylate cyclase activity mediated by the enzymes II of the phosphoenolpyruvate:sugar phosphotransferase system in *Escherichia coli*. *J. Biol. Chem.* **251**, 883–892.

Saier, M. H., Jr., Simoni, R. D., and Roseman, S. (1976b). Sugar transport: Properties of mutant bacteria defective in proteins of the phosphoenolpyruvate:sugar phosphotransferase system. *J. Biol. Chem.* **251**, 6584–6597.

Saier, M. H., Jr., Feucht, B. U., and Mora, W. K. (1977). Sugar phosphate:sugar transphosphorylation and exchange group translocation catalyzed by the enzyme II complexes of the bacterial phosphoenolpyruvate:sugar phosphotransferase system. *J. Biol. Chem.* **252**, 8899–8907.

Saier, M. H., Jr., Straud, H., Massman, L. S., Judice, J. J., Newman, M. J., and Feucht, B. U. (1978a). Permease-specific mutations in *Salmonella typhimurium* and *Escherichia coli* that release the glycerol, maltose, melibiose, and lactose transport systems from regulation by the phosphoenolpyruvate:sugar phosphotransferase system. *J. Bacteriol.* **133**, 1358–1367.

Saier, M. H., Jr., Schmidt, M. R., and Leibowitz, M. (1978b). Cyclic AMP-dependent synthesis of fimbriae in *Salmonella typhimurium*: Effects of *cya* and *pts* mutations. *J. Bacteriol.* **134**, 356–358.

Saier, M. H., Jr., Schmidt, M. R., and Lin, P. (1980). Phosphoryl exchange reaction catalyzed by enzyme I of the bacterial phosphoenolpyruvate:sugar phosphotransferase system. *J. Biol. Chem.* **255**, 8579–8584.

Saier, M. H., Jr., Cox, D. F., Feucht, B. U., and Novotny, M. J. (1982a). Evidence for the functional association of Enzyme I and HPr of the phosphoenolpyruvate–sugar phosphotransferase system with the membrane in sealed vesicles of *Escherichia coli*. *J. Cell. Biochem.* **18**, 231–238.

Saier, M. H., Jr., Keeler, D. K., and Feucht, B. U. (1982b). Physiological desensitization of carbohydrate permeases and adenylate cyclase to regulation by the phosphoenolpyruvate:sugar phosphotransferase system in *Escherichia coli* and *Salmonella typhimurium*. *J. Biol. Chem.* **257**, 2509–2517.

Saier, M. H., Jr., Novotny, M. J., Fuhrman, D. C., Osumi, T., and Desai, J. D. (1983). *In vivo* evidence for cooperative binding of the sugar substrates and the allosteric regulatory protein (Enzyme III[glc] of the phosphotransferase system) to the lactose and melibiose permeases in *Escherichia coli* and *Salmonella typhimurium*. *J. Bacteriol.* **155**, 1351–1359.

Saper, M. A., and Quiocho, F. A. (1983). Leucine, isoleucine, valine-binding protein from *Escherichia coli*. *J. Biol. Chem.* **258**, 11057–11062.

Sarno, M. V., Tenn, L. G., Desai, A., Chin, A. M., Grenier, F. C., and Saier, M. H., Jr. (1984). Genetic evidence for a glucitol-specific enzyme III, an essential phosphocarrier protein of the glucitol phosphotransferase system in *Salmonella typhimurium*. *J. Bacteriol.* **157**, 953–955.

Schiffer, M., and Edmundson, A. (1967). Use of helical wheels to represent the structures of proteins and to identify segments with helical potential. *Biophys. J.* **7**, 121–135.

Schmitt, R. (1968). Analysis of melibiose mutants deficient in α-galactosidase and thiomethylgalactoside permease II in *Escherichia coli* K-12. *J. Bacteriol.* **96**, 462–471.

Scholte, B. J., and Postma, P. W. (1981a). Mutation in the *crp* gene of *Salmonella typhimurium* which interferes with inducer exclusion. *J. Bacteriol.* **141**, 751–757.

Scholte, B. J., and Postma, P. W. (1981b). Competition between two pathways for sugar uptake by the phosphoenolpyruvate-dependent sugar phosphotransferase system in *Salmonella typhimurium*. *Eur. J. Biochem.* **114**, 51–58.

Scholte, B. J., Schuitema, A. R., and Postma, P.W. (1981). Isolation of III^{glc} of the phosphoenolpyruvate-dependent glucose phosphotransferase system of *Salmonella typhimurium*. *J. Bacteriol.* **148**, 257–264.

Scholte, B. J., Schuitema, A. R., and Postma, P. W. (1982). Characterization of factor III^{glc} in catabolite repression-resistant (*crr*) mutants of *Salmonella typhimurium*. *J. Bacteriol.* **149**, 576–586.

Seckler, R., Wright, J. K., and Overath, P. (1983). Peptide-specific antibody locates the COOH terminus of the lactose carrier of *Escherichia coli* on the cytoplasmic side of the plasma membrane. *J. Biol. Chem.* **258**, 10817–10820.

Shapiro, L. (1976). Differentiation in the *Caluobacter* cell cycle. *Annu. Rev. Microbiol.* **30**, 377–407.

Shapiro, L., Nisen, P., and Ely, B. (1981). Genetic analysis of the differentiating bacterium: *Caulobacter crescentus*. *Symp. Soc. Gen. Microbiol.* **31**, 317–339.

Shattuck-Eidens, D. M., and Kadner, R. J. (1983). Molecular cloning of the *uhp* region and evidence for a positive activator for expression of the hexose phosphate transport system of *Escherichia coli*. *J. Bacteriol.* **155**, 1062–1070.

Shuman, H. A. (1982a). The maltose–maltodextrin transport system of *Escherichia coli*. *Ann. Microbiol.* (*Paris*) **133A**, 153–162.

Shuman, H. A. (1982b). Active transport of maltose in *Escherichia coli* K12. *J. Biol. Chem.* **257**, 5455–5461.

Shuman, H. A., and Silhavy, T. J. (1981). Identification of the *malK*-gene product: A peripheral membrane component of the *E. coli* maltose transport system. *J. Biol. Chem.* **256**, 560–562.

Shuman, H. A., Silhavy, T. J., and Beckwith, J. (1980). Labeling proteins with β-galactosidase by gene fusion: Identification of a cytoplasmic membrane component of the *Escherichia coli* maltose transport system. *J. Biol. Chem.* **255**, 168–174.

Simoni, R. D., and Roseman, S. (1973). Sugar transport. VII. Lactose transport in *Staphylococcus aureus*. *J. Biol. Chem.* **248**, 966–976.

Simoni, R. D., Hays, J. B., Nakazawa, T., and Roseman, S. (1973). Sugar transport. VI. Phosphoryl transfer in the lactose phosphotransferase system of *Staphylococcus aureus*. *J. Biol. Chem.* **248**, 957–965.

Steitz, T. A., Ohlendorf, D. H., McKay, D. B., Anderson, W. F., and Matthews, B. W. (1982). Structural similarity in the DNA-binding domains of catabolite gene activator and *cro* repressor proteins. *Proc. Natl. Acad. Sci. U.S.A.* **79**, 3097–3100.

Szmelcman, S., Schwartz, M., Silhavy, T. J., and Boos, W. (1976). Maltose transport in *Escherichia coli*. *Eur. J. Biochem.* **65**, 13–19.

Teather, R. M., Bramhall, J., Riede, I., Wright, J. K., Fürst, M., Aichele, G., Wilhelm, U., and Overath, P. (1980). Lactose carrier protein of *Escherichia coli*: Structure and expression of plasmids carrying the Y gene of the *lac* operon. *Eur. J. Biochem.* **108**, 223–231.

Thompson, J., and Chassey, B. M. (1982). Novel phosphoenolpyruvate-dependent futile cycle in *Streptococcus lactis*: 2-deoxy-D-glucose uncouples energy production from growth. *J. Bacteriol.* **151**, 1454–1465.

Thompson, J., and Chassey, B. M. (1983). Intracellular hexose-6-phosphate : phosphohydrolase from *Streptococcus lactis*: Purification, properties and function. *J. Bacteriol.* **156,** 70–80.

Thompson, J., and Saier, M. H., Jr. (1981). Regulation of methyl-β-D-thiogalacto-pyranoside-6-phosphate accumulation in *Streptococcus lactis* by exclusion and expulsion mechanisms. *J. Bacteriol.* **146,** 885–894.

Thorner, J. W. (1975). Glycerol kinase. *In* "Methods in Enzymology" (W. A. Wood, ed.), Vol. 42, pp. 148–156. Academic Press, New York.

Thorner, J. W., and Paulus, H. (1971). Composition and subunit structure of glycerol kinase from *Escherichia coli. J. Biol. Chem.* **246,** 3885–3894.

Thorner, J. W., and Paulus, H. (1973). Catalytic and allosteric properties of glycerol kinase from *Escherichia coli. J. Biol. Chem.* **248,** 3922–3932.

Tokuda, H., and Kaback, H. R. (1978). Sodium-dependent binding of *p*-nitrophenyl α-D-galactopyranoside to membrane vesicles isolated from *Salmonella typhimurium. Biochemistry* **17,** 698–705.

Tsuchiya, T., and Wilson, T. H. (1978). Cation–sugar cotransport in the melibiose transport system of *Escherichia coli. Membr. Biochem.* **2,** 63–79.

Tsuchiya, T., Lopilato, J., and Wilson, T. H. (1978). Effect of lithium ion on melibiose transport in *Escherichia coli. J. Membr. Biol.* **42,** 45–59.

Tsuchiya, T., Ottina, K., Moriyama, Y., Newman, M. J., and Wilson, T. H. (1982). Solubilization and reconstitution of the melibiose carrier from a plasmid-carrying strain of *Escherichia coli. J. Biol. Chem.* **257,** 5125–5128.

Tsuchiya, T., Oho, M., and Shiota-Niiya, S. (1983). Lithium ion–sugar cotransport via the melibiose transport system in *Escherichia coli. J. Biol. Chem.* **258,** 12765–12767.

Ullman, A., and Danchin, A. (1983). Role of cyclic AMP in bacteria. *Adv. Cyclic Nucleotide Res.* **15,** 1–53.

Viitanen, P., Garcia, M. L., Foster, D. L., Kaczorowski, G. J., and Kaback, H. R. (1983). Mechanism of lactose translocation in proteoliposomes reconstituted with *lac* carrier protein purified from *Escherichia coli*. II. Deuterium solvent isotope effects. *Biochemistry* **22,** 2531–2536.

Vyas, N. K., Vyas, M. N., and Quiocho, F. A. (1983). The 3 Å resolution structure of a D-galactose-binding protein for transport and chemotaxis in *Escherichia coli. Proc. Natl. Acad. Sci. U.S.A.* **80,** 1792–1796.

Wandersman, C., Schwartz, M., and Ferenci, T. (1979). *Escherichia coli* mutants impaired in maltodextrin transport. *J. Bacteriol.* **140,** 1–13.

Wang, J. Y. J., and Koshland, D. E., Jr. (1978). Evidence for protein kinase activities in the prokaryote *Salmonella typhimurium. J. Biol. Chem.* **253,** 7605–7608.

Wang, J. Y. J., and Koshland, D. E., (1981a). The identification of distinct protein kinases and phosphatases in the prokaryote *Salmonella typhimurium. J. Biol. Chem.* **256,** 4640–4648.

Wang, J. Y. J., and Koshland, D. E., Jr. (1981b). Regulatory protein phosphorylation in the prokaryote *Salmonella typhimurium. Cold Spring Harbor Conf. Cell Proliferation* **8,** 637–643.

Wang, J. Y. J., and Koshland, D. E., Jr. (1982). The reversible phosphorylation of isocitrate dehydrogenase of *Salmonella typhimurium. Arch. Biochem. Biophys.* **218,** 59–67.

Wang, J. Y. J., Clegg, D. O., and Koshland, D. E. (1981). Molecular cloning and amplification of the adenylate cyclase gene. *Proc. Natl. Acad. Sci. U.S.A.* **78,** 4684–4688.

Waygood, E. B. (1980). Resolution of the phospho*enol*pyruvate:fructose phosphotransferase system of *Escherichia coli* into two components; enzyme II[fructose] and fructose-induced HPr-like protein (FPr). *Can. J. Biochem.* **58,** 1144–1146.

Waygood, E. B., and Steeves, T. (1980). Enzyme I of the phosphoenolpyruvate:sugar phosphotransferase system of *Escherichia coli*. Purification to homogeneity and some properties. *Can. J. Biochem.* **58,** 40–48.

Waygood, E. B., Meadow, N. D., and Roseman, S. (1979). Modified assay procedures for the phosphotransferase system in enteric bacteria. *Anal. Biochem.* **95,** 293–304.

Weigel, N., Kukuruzinska, M. A., Nakazawa, A., Waygood, E. B., and Roseman, S. (1982a). Sugar transport by the bacterial phosphotransferase system. Isolation and characterization of enzyme I from *Salmonella typhimurium*. *J. Biol. Chem.* **257,** 14461–14469.

Weigel, N., Kukuruzinska, M. A., Nakazawa, A., Waygood, E. B., and Roseman, S. (1982b). Sugar transport by the bacterial phosphotransferase system. Phosphoryl transfer reactions catalyzed by enzyme I of *Salmonella typhimurium*. *J. Biol. Chem.* **257,** 14477–14491.

Weigel, N., Powers, D. A., and Roseman, S. (1982c). Sugar transport by the bacterial phosphotransferase system. Primary structure and active site of a general phosphocarrier protein (HPr) from *Salmonella typhimurium*. *J. Biol. Chem.* **257,** 14499–14509.

Willecke, K., and Pardee, A. B. (1971). Inducible transport of citrate in a grampositive bacterium, *Bacillus subtilis*. *J. Biol. Chem.* **246,** 1032–1040.

Wilson, T. H., and Kusch, M. (1972). A mutant of *Escherichia coli* K12 energyuncoupled for lactose transport. *Biochim. Biophys. Acta* **225,** 786–797.

Wilson, T. H., Kusch, M., and Kashket, E. R. (1970). A mutant in *Escherichia coli* energy-uncoupled for lactose transport, a defect in the lactose operon. *Biochem. Biophys. Res. Commun.* **40,** 1409–1414.

Wright, J. K., Riede, I., and Overath, P. (1981). Lactose carrier protein of *Escherichia coli*: Interaction with galactosides and protons. *Biochemistry* **20,** 6404–6415.

Wright, J. K., Weigel, U., Lustig, A., Bocklage, H., Mieschendahl, M., Müller-Hill, B., and Overath, P. (1983). Does the lactose carrier of *Escherichia coli* function as a monomer? *FEBS Lett.* **162,** 11–15.

Yamada, H., Nogami, T., and Mizushima, S. (1982). Arrangement of bacteriophage lambda receptor protein (*lamB*) in the *Escherichia coli* cell surface. *Ann. Microbiol. (Paris)* **133A,** 43–48.

Yang, J. K., and Epstein, W. (1983). Purification and characterization of adenylate cyclase from *Escherichia coli* K12. *J. Biol. Chem.* **258,** 3750–3758.

Yang, J. K., Bloom, R. W., and Epstein, W. (1979). Catabolite and transient repression in *Escherichia coli* do not require Enzyme I of the phosphotransferase system. *J. Bacteriol.* **138,** 275–279.

Zilberstein, D., Ophir, I. J., Padan, E., and Schuldiner, S. (1982). Na^+ gradientcoupled porters of *Escherichia coli* share a common subunit. *J. Biol. Chem.* **257,** 3692–3696.

Index